Supply Chain Risk Management

Internal Audit and IT Audit

Series Editor: Dan Swanson

Cognitive Hack: The New Battleground in Cybersecurity ... the Human Mind
James Bone
ISBN 978-1-4987-4981-7

The Complete Guide to Cybersecurity Risks and Controls
Anne Kohnke, Dan Shoemaker, and Ken E. Sigler
ISBN 978-1-4987-4054-8

Corporate Defense and the Value Preservation Imperative: Bulletproof Your Corporate Defense Program
Sean Lyons
ISBN 978-1-4987-4228-3

Data Analytics for Internal Auditors
Richard E. Cascarino
ISBN 978-1-4987-3714-2

Ethics and the Internal Auditor's Political Dilemma: Tools and Techniques to Evaluate a Company's Ethical Culture
Lynn Fountain
ISBN 978-1-4987-6780-4

A Guide to the National Initiative for Cybersecurity Education (NICE) Cybersecurity Workforce Framework (2.0)
Dan Shoemaker, Anne Kohnke, and Ken Sigler
ISBN 978-1-4987-3996-2

Implementing Cybersecurity: A Guide to the National Institute of Standards and Technology Risk Management Framework
Anne Kohnke, Ken Sigler, and Dan Shoemaker
ISBN 978-1-4987-8514-3

Internal Audit Practice from A to Z
Patrick Onwura Nzechukwu
ISBN 978-1-4987-4205-4

Leading the Internal Audit Function
Lynn Fountain
ISBN 978-1-4987-3042-6

Mastering the Five Tiers of Audit Competency: The Essence of Effective Auditing
Ann Butera
ISBN 978-1-4987-3849-1

Operational Assessment of IT
Steve Katzman
ISBN 978-1-4987-3768-5

Operational Auditing: Principles and Techniques for a Changing World
Hernan Murdock
ISBN 978-1-4987-4639-7

Practitioner's Guide to Business Impact Analysis
Priti Sikdar
ISBN 978-1-4987-5066-0

Securing an IT Organization through Governance, Risk Management, and Audit
Ken E. Sigler and James L. Rainey, III
ISBN 978-1-4987-3731-9

Security and Auditing of Smart Devices: Managing Proliferation of Confidential Data on Corporate and BYOD Devices
Sajay Rai, Philip Chukwuma, and Richard Cozart
ISBN 978-1-4987-3883-5

Software Quality Assurance: Integrating Testing, Security, and Audit
Abu Sayed Mahfuz
ISBN 978-1-4987-3553-7

Supply Chain Risk Management: Applying Secure Acquisition Principles to Ensure a Trusted Technology Product
Ken Sigler, Dan Shoemaker, and Anne Kohnke
ISBN 978-1-4987-3553-7

Why CISOs Fail: The Missing Link in Security Management—and How to Fix It
Barak Engel
ISBN 978-1-138-19789-3

Supply Chain Risk Management

Applying Secure Acquisition Principles to Ensure a Trusted Technology Product

Ken Sigler, Dan Shoemaker, and Anne Kohnke

CRC Press
Taylor & Francis Group
Boca Raton London New York

CRC Press is an imprint of the
Taylor & Francis Group, an **informa** business

AN AUERBACH BOOK

CRC Press
Taylor & Francis Group
6000 Broken Sound Parkway NW, Suite 300
Boca Raton, FL 33487-2742

International Standard Book Number-13: 978-1-138-19735-0 (Hardback)
International Standard Book Number-13: 978-1-138-19733-6 (Paperback)

Library of Congress Cataloging-in-Publication Data

Names: Sigler, Kenneth, author. | Shoemaker, Dan, author. |
Kohnke, Anne, author.
Title: Supply chain risk management : applying secure acquisition principles
to ensure a trusted technology product / Ken Sigler, Dan Shoemaker,
Anne Kohnke.
Description: New York : CRC Press, [2018] | Series: Internal audit and IT audit
Identifiers: LCCN 2017030801 | ISBN 9781138197350 (hb : alk. paper) |
ISBN 9781138197336 (pb : alk. paper) | ISBN 9781315279572 (e)
Subjects: LCSH: Business logistics. | Risk management. | Data protection. |
Computer networks--Security measures.
Classification: LCC HD38.5 .K64 2018 | DDC 658.7--dc23
LC record available at https://lccn.loc.gov/2017030801

Visit the Taylor & Francis Web site at
http://www.taylorandfrancis.com

and the CRC Press Web site at
http://www.crcpress.com

Contents

Foreword xi
Preface xiii
Authors xvii
Contributions xix
Chapter Structure and Summary xxi

1   **Why Secure Information and Communication Technology Product Acquisition Matters** 1
    Introduction to the Book 1
    Underwriting Trust and Competence 2
    Justification and Objectives of the Book 3
    The Five-Part Problem 4
    Putting Product Assurance into Practice 7
    The Supply Chain and the Weakest Link 8
    Visibility and Control 9
    Building Visibility into the Acquisition Process 11
    The Seven Phases of ICT Acquisition Practice 13
          Practice Area One: Procurement Program Initiation and Planning 14
          Practice Area Two: Product Requirements Communication
             and Bidding 16
          Practice Area Three: Source Selection and Contracting 16
          Practice Area Four: Supplier Considerations 20
          Practice Area Five: Customer Agreement Monitoring 21
          Practice Area Six: Product Acceptance 22
          Practice Area Seven: Project Closure 23
    Building the Foundation: The Role of Governance in Securing the
        ICT Supply Chain 23
    The Use of Standard Models of Best Practice 32
    Chapter Summary 33
    Key Concepts 38
    Key Terms 39
    References 40

2 Building a Standard Acquisition Infrastructure 41
ISO/IEC 12207 42
Agreement Processes: Overview..45
Acquisition Process...47
Acquisition Activity: Acquisition Preparation 50
Concept of Need 51
Define, Analyze, and Document System Requirements 52
Consideration for Acquiring System Requirements 53
Preparation and Execution of the Acquisition Plan...................................54
Acceptance Strategy Definition and Documentation 55
Prepare Acquisition Requirements...56
    Acquisition Activity: Acquisition Advertisement 57
    Acquisition Activity: Supplier Selection 58
    Acquisition Activity: Contract Agreement.....................................59
    Acquisition Activity: Agreement Monitoring..................................60
    Acquisition Activity: Closure 61
Supply Process...61
    Supply Activity: Opportunity Identification....................................63
    Supply Activity: Supplier Tendering..63
    Supply Activity: Contract Agreement...65
    Supply Activity: Contract Execution 67
    Supply Activity: Product/Service Delivery and Support..................74
    Supply Activity: Closure 75
Chapter Summary..75
Key Terms 76
References 77

3 The Three Building Blocks for Creating Communities of Trust 79
Introduction to Product Trust 79
Building a Basis for Trust 81
The Hierarchy of Sourced Products 82
The Problem with Sourced Products..88
Promoting Trust through Best Practice 92
Moving the Product up the Supply Chain 93
The Standard Approach to Identifying and Controlling Risk.....................95
The Three Standard Supply Chain Roles 96
    The Acquirer Role...97
    The Supplier Role 101
    The Integrator Role..104
Information and Communication Technology Product Assurance............105
Adopting a Proactive Approach to Risk 107
People, the Weakest Link 108

Chapter Summary 110
Key Concepts 114
Key Terms 115
References 115

**4 Risk Management in the Information and Communication
Technology (ICT) Product Chain 117**
Introduction.. 117
Supply Chain Security Control Categorization................................... 119
Categorization Success through Collaboration 123
Supply Chain Security Control Selection 124
The Eight Tasks of Control Selection..128
    Documentation Prior to Selection...128
    Select Initial Security Control Baselines and Minimum
        Assurance Requirements...128
    Determine Need for Compensating Controls...............................131
    Determine Organizational Parameters...132
    Supplement Security Controls 132
    Determine Assurance Measures for Minimum Assurance
        Requirements 134
    Complete Security Plan 135
    Develop a Continuous Monitoring Strategy................................136
Supply Chain Security Control Implementation 137
    Implement the Security Controls Specified in the Security Plan........138
Security Control Documentation 141
Supply Chain Security Control Assessment 142
The Four Tasks of Security Control Assessment..................................144
Implications of Security Control Authorization to the Supply Chain 149
The Four Tasks of Security Control Authorization.............................. 151
Supply Chain Risk Continuous Monitoring....................................... 155
The Seven Tasks of Security Continuous Monitoring 157
    Determine the Security Impact of Changes................................. 158
    Assess Selected Security Controls 159
    Conduct Remediation Actions.. 159
    Update the Security Plan, Security Assessment Report, and
        POA&M...160
    Report the Security Status...160
    Review the Reported Security Status on an Ongoing Basis............. 161
    Implement an ICT System Decommissioning Strategy 162
Chapter Summary.. 162
Key Terms 164
References 165

5 Establishing a Substantive Control Process 167
Introduction: Using Formal Models to Build Practical Processes 167
Why Formal Models Are Useful 169
NIST SP 800-161, Supply Chain Risk Management Practices for
     Federal Information Systems 170
The 21 Principles for SCRM 172
     Principle 1: Maximize Acquirer's Visibility into the Actions of
          Integrators and Suppliers in the Process 173
     Principle 2: Ensure That the Uses of Individual Supply Chain
          Components Are Kept Confidential 174
     Principle 3: Incorporate Conditions for Supply Chain Assurance
          in Specifications of Requirements 175
     Principle 4: Select Trustworthy Elements and Components 176
     Principle 5: Enable a Diverse Supply Chain—Do Not
          Sole Source 176
     Principle 6: Identify and Protect Critical Processes and Elements 176
     Principle 7: Use Defensive Design in Component Development 176
     Principle 8: Protect the Contextual Supply Chain Environment 177
     Principle 9: Configure Supply Chain Elements to Limit
          Access and Exposure 177
     Principle 10: Formalize Service/Maintenance Agreements 177
     Principle 11: Test throughout the SDCL 178
     Principle 12: Manage All Pertinent Versions of
          the Configuration 178
     Principle 13: Factor Personnel Considerations into Supply
          Chain Management 179
     Principle 14: Promote Awareness, Educate, and Train Personnel
          on Supply Chain Risk 179
     Principle 15: Harden Supply Chain Delivery Mechanisms 179
     Principle 16: Protect/Monitor/Audit the Operational Supply
          Chain System 180
     Principle 17: Negotiate and Manage Requirements Changes 180
     Principle 18: Manage Identified Supply Chain Vulnerabilities 181
     Principle 19: Reduce Supply Chain Risks during Software
          Updates and Patches 181
     Principle 20: Respond to Supply Chain Incidents 181
     Principle 21: Reduce Supply Chain Risks during Disposal 182
Making Control Structures Concrete: FIPS 200 and
     NIST 800-53(Rev 4) 182
Application of FIPS 200 and NIST 800-53(Rev 4) to Control
     Formulation 183
The Generic Security Control Set 186

NIST 800-53 Control Baselines 186
Detail of Controls 187
Six Feasibility Considerations for NIST 800-53 188
NIST 800-53 Catalog of Baseline Controls 190
Implementing Management Control Using the Standard
    NIST SP 800-53 Rev. 4 Control Set 191
Practical Security Control Architectures 192
Control Statements 192
Supplemental Guidance 193
Control Enhancements 193
Real-World Control Formulation and Implementation 193
Limitations of the 800-53 Approach in SCRM 194
Chapter Summary 196
Key Concepts 199
Key Terms 200
References 201

6 Control Sustainment and Operational Assurance..............................203
Sustaining Long-Term Product Trust 203
Step 1: Establish and Maintain Situational Awareness 205
Step 2: Analyze Reported Vulnerability and Understand
    Operational Impacts 209
    Environmental Monitoring 210
    Vulnerability Reporting 210
    Vulnerability Response Management 211
Step 3: Obtain Management Authorization to Remediate 212
    Understand Impacts 213
    Communicating with Authorization Decision-Makers 215
Step 4: Manage and Oversee the Authorized Response 216
    Responding to Known Vulnerabilities with Fixes 217
    Responding to Known Vulnerabilities without Fixes 217
    Fixing an Identified ICT Supply Chain Vulnerability .... ■ 218
Step 5: Evaluate the Correctness and Effectiveness of the
    Implemented Response 219
Step 6: Assure the Integration of the Response into the Larger
    Supply Chain Process 223
Establishing a Supply Chain Assurance Infrastructure 225
    Policies for Operational Assurance: Method, Measurement,
        and Metrics 226
Building a Practical Supply Chain Sustainment Function 228
Generic Management Roles 230
Conducting the Day-to-Day Operational Response Process 230

Response Management Process Planning.................................231
Deciding What to Secure 232
Enforcing Management Control 232
Status Assessment...233
Maintaining Documentation Integrity 234
Chapter Summary...234
Key Concepts..237
Key Terms 237
References 238

7 Building a Capable Supply Chain Operation....................239
Introduction...239
Why a Capability Maturity Model?.................................241
A Staged Model for Increasing Capability in Supply
    Chain Management 242
Level One: The Initial Level 244
Level Two: The Repeatable Level 244
    Level Two: Acquisition Planning................................ 246
    Level Two: Solicitation...247
    Level Two: Requirements Development and Management...........248
    Level Two: Project Management 249
    Level Two: Contract Tracking and Oversight 250
    Level Two: Evaluation 251
    Level Two: Transition to Support 251
Level Three: The Defined Level 253
    Level Three: Process Definition and Maintenance 254
    Level Three: User Requirements.................................256
    Level Three: Project Performance Management.................257
    Level Three: Contract Performance Management..............257
    Level Three: Acquisition Risk Management 258
    Level Three: Training Program Management..................259
Level Four: The Quantitative Level.................................. 260
    Level Four: Quantitative Process Management................. 260
    Level Four: Quantitative Acquisition Management 261
Level Five: The Optimizing Level 262
    Level Five: Continuous Process Improvement 262
    Level Five: Acquisition Innovation Management..............263
Practical Evaluation of Supply Chain Process Maturity............ 264
Maturity Rating Schemes 266
Chapter Summary...267
Key Terms 272
References 272

Index 273

Foreword

Complexities in the cyber supply chain have introduced new avenues for exploitation and manipulation attracting numerous U.S. adversaries.

Megan Mance
Cyber Supply Chain Security and Potential Vulnerabilities within U.S. Government Networks (June 15, 2016)

On November 19, 2015 Admiral Michael Rogers, the Chief of the National Security Agency and U.S. Cyber Command, told an audience that his principal concern is data manipulation through network intrusion. U.S. adversaries are continually looking for new and creative ways to gain access to U.S. government networks. One specific cyber threat that deserves greater attention is the *cyber supply chain security* within the federal government and the vital role of government contractors in this area.

The 2017 U.S. Presidential Executive Order on Strengthening the Cybersecurity of Federal Networks and Critical Infrastructure (dated May 11, 2017) calls for the U.S. government departments and agencies to report "on cybersecurity risks facing the defense industrial base, including its *supply chain*, and United States military platforms, systems, networks, and capabilities, and recommendations for mitigating these risks."

Even a trusted supplier can unwittingly integrate components that might be obtained from untrustworthy sources. Unfortunately, this is the likely situation given the globally sourced commercial off-the-shelf (COTS) strategies popular in our current national security and information and communication technology (ICT) business organizations.

It is too difficult to provide across-the-board assurance for all tiers/levels of all products, because most products are agilely sourced in a global environment; if we cannot do it all, the key concept here is to take a risk-based approach to secure what we can, to at least provide across-the-board assurance for some critical capabilities. Therefore, it will be necessary for enterprises to identify their most critical capabilities or functions, and for those mission-essential functions, they will seek to ensure trustworthiness of every product, component, and subcomponent enabling that capability/function.

Planning for the tracking of subcomponents and components brought together in a select product or system is required, because it would be nearly impossible to trace back after the fact. So, a big part of ensuring a trustworthy sourcing process rests on the ability of the supplier to prove that they can deliver the knowledge of their supply chain processes: what is planned and how they will manage deviations from predicted plans. This type of detailed supply chain planning (SCRM) also leads to on-cost, timely delivery of these trusted components. This is an especially difficult requirement with complex technology projects, due to their layers of design complexity and a multitiered global supply chain.

Currently, given this complexity of most ICT projects, it is difficult for any supplier to provide this sort of assurance/supply chain guarantee. Part of the problem is that until recently there had been no defined requirement for this type of process/ system or any adequate description of what it takes to plan and provide end-to-end technology supply chain assurance. That has changed recently, primarily due to the recognition that elements of our critical infrastructure and national security systems may already contain poor quality and/or malicious items placed through insecure supply chains and slipshod sourcing.

Currently, efforts are increasing to address problems with embedded malware and counterfeits due to supply chain breakdowns and ultimately enabling compromised functional capabilities. These efforts are driving a need for supply chain assurance for select systems/capabilities. Coordination of this complex work requires a common and coherent set of control processes and activities, which will allow managers to understand the precise security status of any given component as it moves through design, manufacture, final product integration, testing, and assurance of a well-documented build of materials (BOM), which become well-maintained and dynamic BOM, due to strong configuration management for both hardware and software, throughout the life cycle. In that respect, an authoritative, mutually agreed-upon process for independently assuring organizational trust in sourced products becomes a necessity.

This is not a common situation today because of the distances and global elements of business. The ideal would be a well-defined (risk-based) process, with agreed-upon taxonomy, to evaluate and verify trust at all levels up and down the supply chain, to ensure that control is maintained through a formal and disciplined process. This overall control framework with well-defined processes, activities, and tasks has a good start in NIST 800-161 standards, practices, and controls for supply chain assurance. This is a critically enabling first step in the process of assuring globally sourced products, enabling trusted mission essential functions, through ICT SCRM.

Donald R. Davidson Jr.
Director, Cybersecurity (CS) Risk Management

Preface

Today's information and communication technology (ICT) organizations increasingly find themselves relying on others for their success. Historically, medium- and large-size organizations have spent less than a third of their budgets on purchased goods and services, having relied on internal sources for these. Today, those same organizations spend most of their budget on purchased commercial off-the-shelf (COTS) goods and services. This is in large part because of the advantages ICT organizations have found in strategies such as globalization, outsourcing, supply-base rationalization, just-in-time deliveries, and lean inventories. Additionally, many companies have consolidated operations both internally and externally to achieve economies of scale.

While globalization, extended supply chains, and supplier consolidation offer many benefits in efficiency and effectiveness, they can also make supply chains more brittle and can increase information security risks that can lead to supply chain disruption. Historic and recent events have proven the need to identify and mitigate such risks. Recent political accusations have shown how security breaches can extend well beyond domestic boundaries and interfere with international trade and disrupt many elements of global supply chains, including supply, distribution, and communications. In extreme cases, a single security breach at one location can severely interfere with the capabilities of an organization.

Effective supply chain risk management (SCRM) is essential to a successful business. It is a competence and capability many enterprises have yet to develop. In some areas, both problems and practices are well defined. In others, problems are defined, but practices are developing. In still other areas, both the definition of the problems and the practices needed to address them are developing. In sum, SCRM is an evolving field.

It is important to note that ICT security risk cannot be eliminated. Because of its complex nature, various tools can be used to give organizations and governments the ability to build up an overall picture of the risk situation and plan a mitigation strategy to address critical areas. Likewise, with the growing emphasis on globalization, business process outsourcing, and the need to control terrorism, there is a greater need to understand and handle supply chain vulnerabilities throughout the entire life cycle from agreement to procurement and operation.

Guidelines for managing information security risk were developed by the National Institute of Standards and Technology (NIST) in March of 2011; they propose risk management practices at all levels of the organization and should be followed to facilitate adequate risk assessment, response, and monitoring (NIST SP 800-39). In 2015, NIST expanded the scope of the guidelines in the development of new guidelines (NIST SP 800-161) that address the ever-increasing implications of SCRM while incorporating the requirements of specific management, operational, and technical ICT security controls under the Federal Information Security Management Act. Until now, however, there has not existed an "easy-to-understand" approach for implementing NIST SP 800-161 parallel to the definitions of other international standards, such as ISO/IEC 12207, for the purpose of providing security mechanisms within end-to-end ICT supply chain agreement, procurement, and operation that lead to a comprehensive capacity maturity model.

This book is based upon the belief that the acquisition process is a strategic planning and governance concern. The solution is a formally defined and implemented infrastructure of best practices aimed at specifically optimizing the coordination and control of the acquisition process across the organization. As with any complex process deployment, this can only be substantiated through a rational and explicit framework of auditable procedures. The creation and deployment of those procedures is at the core of what is being presented.

One of the underlying premises of this book is to detail the reasons why formal organizational processes and methods for acquiring secure products are valuable. You will see how fundamental security activities provide the basis for the most effective assurance of the technology used. You will also discover the importance of expert advice concerning the best practices for building these formal processes. Since continuous capability improvement is the essence of maintaining an effective security posture, we will describe a maturity model-based approach to acquisition process improvement.

Who Should Read This Book?

This book will provide a valuable insight to anyone who acquires technology products. This includes COTS and government off-the-shelf products as well as the participants in the supply chain at any level. However, given the management focus, it would be particularly useful to process architects and higher level executives responsible for assurance of the technology infrastructure.

This book can also serve as a good general knowledge text for general interest practitioners. Since the ideas have practical business application, they seem highly attractive to any manager responsible for acquiring any form of complex products. The inclusion of a maturity model in Chapter 7 makes it especially attractive to strategic planners and any other type of security policy manager or upper level strategic decision-maker.

In a very detailed and organized fashion, this book presents the concepts of secure acquisition and ICT SCRM operations as an all-in-one concept. As such, there is no assumption about specialized knowledge. You will learn how to create a systematic and secure acquisition process as well as how to create a risk-based control structure for all levels of the supply chain. You will learn how to establish systematic sustainment and reporting within this structure and how to increase its capability.

This revolves around the steps to define the standard processes, activities, and tasks for the customer–supplier relationship, the attendant control objectives, and the auditing and reporting systems for the supply chain. Guidance for carrying this out is supported by expert standards of best practice, which are commonly accepted in the field and easily understandable.

At the end of this book, you will be able to

1. Implement a formal, organization-wide, standards-based trust in sourced products.
2. Define a comprehensive control structure to ensure continuous assurance.
3. Create a standard process to achieve higher stages of requisite capability.

Authors

Ken Sigler has been a faculty member of the Computer Information Systems (CIS) program at the Auburn Hills Campus of Oakland Community College in Michigan since 2001. His primary research is in the areas of software management, software assurance, and cybersecurity. He originally developed the college's CIS program option entitled "Information Technologies for Homeland Security" and correlated the relationship between that program and the Committee for National Security Standards of the National Security Agency. Mr. Sigler serves on the board of directors of the Colloquium for Information System Security Education (CISSE) and represents his college as the liaison to the Midwest Chapter for CISSE. Throughout his tenure at the college, he has also served as post-secondary liaison to the articulations program with Oakland County Michigan secondary school districts. Through that role, he developed a 2+2+2 Information Security Education process leading students through information security coursework at the secondary level into a 4-year articulated program, leading to a career in information security at a federal agency. Mr. Sigler is a member of the University of Detroit Mercy Center for Cybersecurity & Intelligence Studies Board of Advisors, Institute of Electrical and Electronics Engineers, Distributed Management Task Force, and Association for Information Systems.

Dan Shoemaker, PhD, is principal investigator and senior research scientist at the University of Detroit Mercy (UDM)'s Center for Cyber Security and Intelligence Studies. Mr. Shoemaker has served 30 years as a professor at the UDM with 25 of those years as department chair. He served as a cochair for both the Workforce Training and Education and the Software and Supply Chain Assurance Initiatives for the Department of Homeland Security and was a subject-matter expert for the NICE Workforce Framework 2.0. Mr. Shoemaker has coauthored six books in the field of cybersecurity and has authored more than 100 journal publications. He earned his PhD from the University of Michigan, Ann Arbor, Michigan.

Anne Kohnke, PhD, is an assistant professor of information technology (IT) at Lawrence Technological University, Southfield, Michigan. After a 25-year career in IT, Anne transitioned from a vice president of IT and chief information security officer (CISO) position into full-time academia in 2011. Anne's research is focused in the areas of cybersecurity; risk management; threat modeling; and user's attitudes, decision making, and comprehension of the risks of installing Android mobile applications. Anne has coauthored four books in the field of cybersecurity and earned her Ph.D. from Benedictine University.

This text is one of several titles these three authors have written on the topics of cybersecurity, risk management, and security controls within the Taylor & Francis Internal Audit and IT Audit Series. Other titles include *The Complete Guide to Cybersecurity Risks and Controls* (2016), *A Guide to the National Initiative for Cybersecurity Education (NICE) Cybersecurity Workforce Framework 2.0* (2016), and *Implementing Cybersecurity—A Guide to the National Institute of Standards and Technology Risk Management Framework* (2017).

Contributions

The collaborative work of the authors would not be successful without the organizational assurance and support of Tamara Shoemaker, who continues to be a solid rock for each of the authors and thoroughly enjoyable to work with. In addition to her role, affectionately referred to as *The Boss*, Shoemaker is the director of the University of Detroit Mercy's Center for Cybersecurity and Intelligence Studies. Additionally, Shoemaker serves in the capacity of operations manager for the Colloquium for Information Systems Security Education (CISSE).

None of our titles would be successful without the continued guidance and support of our acquiring editor, Rich O'Hanley, and the lead to the Internal Audit and IT Audit Series, Dan Swanson. Much thanks is also extended to the project management and editorial staff that helped bring this book to successful publication.

Chapter Structure and Summary

Chapter 1: Why Secure Information and Communication Technology Product Acquisition Matters

The goal of this chapter is to demonstrate how a formal approach to acquisition security can be used to ensure the integrity of the technology base of an organization. The key concept here is "across-the-board trust." Because ALL of the potential components of the technology base are involved in the secure functioning of the system, every aspect of that base must be trustworthy. The reader will discover how formal processes and a standard point of reference are necessary to establish adequate trust.

Chapter 2: Building a Standard Acquisition Infrastructure

Two standards are relevant to the definition of a robust acquisition assurance infrastructure. At the concept level, this is the customer–supplier process defined in the "Agreement" processes of the ISO/IEC 12207 Standard. This standard has been widely accepted for over 20 years as the authoritative definition of what, at a minimum, must be undertaken to achieve proper technology acquisition. These standard recommendations can then be tailored into a specific process for any given organizational application. Thus, the need for a single, fully defined infrastructure is a precondition for the definition of the body of knowledge for secure supply chain risk management. As such, the remainder of this book will outline the means to specifically implement the recommendations of the NIST IR 800-161 model within the larger ISO/IEC 12207 Agreement process. The aim is to detail how these explicit recommendations for customer, integrator, and supplier performance fit and work with the 12207 requirements for proper customer–supplier relationships.

Chapter 3: The Three Building Blocks for Creating Communities of Trust

In this chapter, you will learn why a formal, comprehensive, standards-based definition of the activities and tasks necessary to ensure trust is critical to the process. The elements of the product supply chain are hard to identify, let alone ensure. Due to its layers of complexity, this is a difficult task to perform with complex technology development and integration projects, particularly given the fact that most products are integrated up a multilevel supply chain that is often offshore based.

The aim of this chapter is to give an overview of the only existing standard framework for the practice of comprehensive control over complex builds. Most products are developed in multilayered, multivendor, and even multicultural team settings. In order to ensure trust, all of this must be fully coordinated and controlled up and down the supply chain. Coordination of this degree of complex work requires a common and coherent control process and control activities, which will allow managers to understand the exact security status of any given component as it moves to final product integration, testing, and assurance.

Chapter 4: Risk Management in the ICT Product Chain

The process of risk management (identifying and controlling information as it is created within the supply chain), risk identification (examining, documenting, and assessing the security concerns represented by a given component within the supply chain), and risk control (applying controls to reduce identified risks), as well as prioritizing its importance will be described here. It is hard to ensure against threats to the components of an evolving product because the development process is normally dispersed across a number of organizations at various levels of integration. That is potentially risky because any breach of the product development chain can compromise the entire product. The term "weakest link" applies here. Also, there is the issue of offshore development of COTS products. Work across organizational boundaries as defined by agreement is the basic approach to the development of most complex technology products. But most of these relationships are undefined. Software in particular is intangible and dynamically changeable. Thus, it is almost impossible to get an exact understanding of product status as it moves up the development chain. Consequently, explicit and trustworthy risk control processes have to be applied at all levels of the supply chain.

Chapter 5: Establishing a Substantive Control Process

It might seem a little simplistic to say that the problem with developing any complex technology is that it is too complex. But the fact is that control must be established at all levels up and down the supply chain in order to be able to say with certainty that the product can be trusted. The only way to ensure that control is through a formal and disciplined process of assurance. This is the role of a formally constituted and organizationally sanctioned set of processes, activities, and tasks, which have been formulated into a standard acquisition control structure. The problem lies in knowing exactly what constitutes the elements of proper behavior. Thus, this text will present the only existing standard recommendations for the activities needed to ensure the acquisition process. This includes the description of the overall control framework itself as well as the processes, activities, and tasks that the organization must undertake to establish actionable behaviors that can be audited for compliance with the recommendations of standard best practice. In this respect, the ISO 12207 Agreement process will be mapped to the recommendations of NIST 800-161 in order to describe a top-to-bottom concept of secure acquisition assurance. The aim is to help you understand how to establish a standard and auditable secure acquisition process. This includes methods for initiating, planning, executing, and following up/remediating active behaviors for the purposes of systematic control. It includes the definition and assignment of all roles and responsibilities for every participant in the supply chain—customer, supplier, and integrator—and the best practices for documentation and reporting of control information to appropriate sources.

Chapter 6: Control Sustainment and Operational Assurance

The only way to ensure proper implementation of a critical process is through the routine operational sustainment of the active controls that constitute it. This in essence involves tailoring, deploying, and validating a suitable set of behavioral controls and then monitoring their integrity and effectiveness throughout the life cycle of the acquisition process. Basic steps must be carried out to ensure systematic integrity no matter what the actual situation might be. It is necessary to validate the selected control set to assure the effectiveness as well as confirm the accuracy of the defensive scheme. Thus, it is necessary to conduct regular monitoring testing and analysis of the complete set of acquisition assurance activities to understand its status and functioning. This includes steps to detect any malfunctioning within the control set and procedures to ensure that subsequent corrective action will be undertaken.

Sustainment operations begin after the acquisition process is operationally deployed. The sustainment process is planned, implemented, and monitored in the same fashion as any other organizational-level activity. It normally embodies the criteria and factors for judging success. The intention is to be able to say with assurance that the aggregate controls for any given acquisition are effective given the aims of the organization. Operationally, this should take place within a defined reporting and decision-making structure. Because the overall purpose of assurance is to produce a trustworthy assurance outcome, the outcome of sustainment is continuous assurance of process correctness.

Chapter 7: Building a Capable Supply Chain Operation

The role of any form of assurance process is to ensure continuing confidence in the products that are being acquired. However, since managers do not actually do the work, and the product is normally too complex to understand anyway, the organization has to adopt and utilize some form of standard control process in order to ensure product integrity. A capability-based process ensures that reliability and integrity are designed for and built into the products in the first place rather than added on at the end.

The assurance of the proper functioning of the control process is what actually certifies the correctness of the product. In that respect, the aim of all technology assurance activity is to ensure the continuous trustworthy and reliable functioning of all of the deployed controls. Process capability improvement provides a given organization with a template for continuous adaptation and improvement. The assumption is that a technology management system that is based on and follows a commonly accepted model of best practice ensures best-of-breed acquisition assurance. The problem is how to get there. The objective of this chapter is to provide a standard model for capability maturity development for any organization. The assumption is that capability is attained in easy-to-accomplish stages rather than in one impossible leap. Capability maturity models have been utilized in a number of high-technology settings for years. Their general form is well understood and adaptable to the standard practices we are discussing here. Thus, we will specify and describe what needs to take place in a practical sense in order to implement such an approach to acquisition security.

Chapter 1

Why Secure Information and Communication Technology Product Acquisition Matters

At the conclusion of this chapter, the reader will understand the following (Figure 1.1):

- The role and importance of a formal sourcing process in ensuring organizational security
- The standard elements of acquisition management practice
- The concerns and issues associated with insecure supply chains
- The general structure and principles of the ICT supply chain risk management (SCRM) process
- The nine large elements of formal ICT SCRM
- The role and importance of standard models of best practice

Introduction to the Book

The purpose of this book is to ensure an understanding of the strategic process of trusted product acquisition, which is directly associated with the discipline of SCRM. This chapter will introduce the concepts and principles of formal trusted

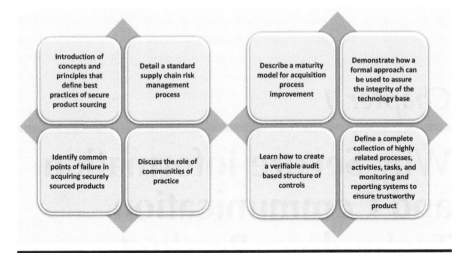

Figure 1.1 Objectives of the book.

product acquisition governance as well as the standard principles and underlying activities that define best practice in the performance of secure product sourcing.

This book will also detail a standard SCRM process that is integral to securing ICT acquisition in a global business environment. It will identify the common points of failure in acquiring adequately secure sourced products, and it will explain the factors that drive those failures. Readers will see how difficult it is to acquire ICT products that are trustworthy and secure, and they will understand the fundamental causes of that difficulty. Readers will discover the role of communities of practice in the overall process of building a complex ICT product, and since continuous capability improvement is the essence of maintaining an effective security posture, Chapter 7 will describe a maturity model for Acquisition Process Improvement.

The goal of this chapter is to demonstrate how a formal approach to acquisition and supply chain security can be used to assure the integrity of the technology base of any organization. The key concept here is "across-the-board trust" because *all* of the potential components of the technology base are involved in the secure functioning of the system. Consequently, every aspect of that base must be dependably secure or "the weakest link" applies. The reader will discover how formal processes and a standard baseline reference are necessary to establish that requisite level of trust.

Underwriting Trust and Competence

The vast range of ICTs have created our digital culture. Consider that 30 years ago you could not shop, bank, buy stocks online, play games, or interact with people on a mobile device. Now that is all possible, and new opportunities seem to pop up at an unthinkably frantic rate. At the same time, because of the dependence on the

Internet, it is critically important to be able to trust the security and integrity of all of our ICT products, and that is demonstrably not the case.

According to the Privacy Rights Clearing House, close to one billion consumer records have been lost or stolen over the last decade. According to McAfee and the Center for Strategic and International Studies, that translates to $300 billion to $1 trillion in annual loss. Therefore, it is not surprising that industry and government have decided to address the problem of ICT product security. Just like buying a suit off the rack rather than having it bespoke tailored means that the customer will get it faster and cheaper, the business logic makes it inescapable for most modern companies to purchase rather than develop their own ICT products. Businesses want their solutions now, not at some time in the indeterminate future, and they do not want to spend the R&D money to back the development of custom packages. In many respects, because solutions are purchased rather than built, the procurement staff is as critical to the security of the ICT operation as the technical staff.

An organization's ICT procurement process is no different from any other purchasing function in that the purpose of any procurement activity is to acquire an effective product for the organization. Consequently, whether the product is a video game or a piece of sophisticated military hardware, the activities that take place within the acquisition process have to be logically related, controlled, and coordinated. A standard model of the best practices to be carried out within that process simply ensures that the control is implemented systematically and is effectively maintained through the specific actions of the individuals who are responsible for performing the assigned task.

From a security and integrity standpoint, what this implies is that every individual action in the overall process has to be rationally and properly placed in the timeline for execution. Additionally, each task must be fully and correctly integrated into the overall activity. Therefore, at its core, the acquisition process that will be discussed here must be well defined and properly executed. It must ensure that proper relationships are maintained among the larger set of actions that have been arrayed to achieve a given purpose.

Justification and Objectives of the Book

Perhaps the best way to justify this book is the statement that it has been long overdue. Technology systems are complex and their elements are indistinguishable by normal inspection. Thus, the usual way to acquire trustworthy ICT products has been to only deal with suppliers who are "known and trusted" over a reasonable period of time. Even so, in a modern global sourcing environment, a trusted supplier has the potential to integrate subcomponents that are obtained from untrustworthy sources into a system. Therefore, when it comes to acquiring the technology needed, any purchaser of an ICT product is essentially "buying a pig in a poke," so to speak. This is a particularly egregious situation given the "faster-cheaper-better" mentality of current companies, and it has led to an overreliance on suppliers' commercial-off-the-shelf (COTS) system security to leverage development strategies.

The problem is that until recently there has been no common body of knowledge that can be relied on to provide a standard set of practices for executing secure, end-to-end technology purchases. Fortunately, this has changed primarily due to the dawning recognition that elements of our critical infrastructure may already contain malicious items, which have been placed there as a result of insecure supply chains and a slipshod open-source acquisition process.

The ideas presented in this book are well-established aspects of a single process that has been developed and promulgated by the federal government to ensure trusted product acquisition in its particular space. Specifically, this book presents the concepts of ICT SCRM from the perspective of NIST SP 800-161, which is the first standard body of knowledge for secure SCRM (NIST, 2015). In this book, you will learn how to create a verifiable audit-based structure of controls, which will ensure comprehensive security for all types of sourced ICT products. We will explain how to establish systematic security within the supply chain as well as how to build auditable trust into the products and services that are acquired by the organization.

In addition, we will detail a unique capability maturity development process that will help foster an increasingly competent process. The overall aim of this book is to define a complete and correct collection of highly related processes, activities, and tasks as well as the attendant monitoring and reporting systems to ensure a trustworthy product. A practical and standard means of leveraging the acquisition process to higher levels of capability maturity is also explained in this text. The details of this process are captured in a very well-known and widely accepted approach to capability maturity development. Thus, the information in this book is both authoritative and commonly agreed upon.

The Five-Part Problem

As we said in the last section, this book centers on the belief that SCRM is a strategic governance concern. Thus, the practical governance solution to the acquisition process is a formally defined and concrete infrastructure of best practices, which are aimed at ensuring sufficient coordination and control over the entire process. The objective is to ensure that all sourced products fall within certain levels of trust. As with any complex goal, the assurance of product trustworthiness can only be substantiated through activities that take place within a rational and explicit framework of auditable procedures. Thus, the basis for creating and deploying these procedures is presented in this chapter.

The General Accounting Office (GAO) summarized the concerns associated with organizational ICT SCRM in a March 23, 2012, report. ICT risk issues fall into five categories, each of which has a slightly different implication for product integrity: "installation of malicious logic on hardware or software; installation of counterfeit hardware or software; failure or disruption in the production or distribution of a critical product or service; reliance upon a malicious or unqualified service provider for the performance of a technical service; and installation of unintentional vulnerabilities on software or hardware" (GAO, 2012, p. 1) (Figure 1.2).

Malicious logic is embedded in a product to fulfill some specific purposes. Malicious objects are by definition not part of the intended functionality; therefore, in order to find and eliminate any instance, rigorous testing and inspection is required. Embedding a malicious object in a product is always a hostile act, and assurance that a product is free of malicious code should be a high priority with any ICT customer. Nonetheless, since it is hard enough to ensure the quality and security of the functions that *ought* to be present in a piece of software, it is asking a lot to expect that functions that should *not* be present should also be identified and eliminated. Therefore, it is almost impossible to estimate how much malicious code currently resides in ICT products. Because the decision to embed a piece of malicious logic in a product is intentional, one of the most effective ways to ensure against the presence of such objects is to maintain strict oversight and control over ICT development, sustainment, and acquisition work.

Counterfeits are not just an acquisition issue. Counterfeit parts can appear at any stage in the development and sustainment of ICT products. Counterfeits execute product functions as intended and threaten product security and integrity because they are not the same as the actual part. Generally, the purpose of a

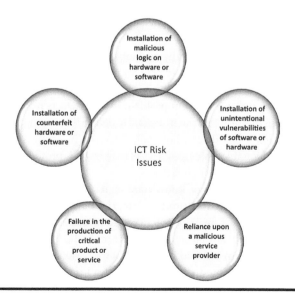

Figure 1.2   ICT risk issues.

counterfeit is to save money or supply a feature that the maker is otherwise incapable of providing. As a result, counterfeits embody shortcuts in product quality or security that can fail in many ways. Because they function like the original part, it is often hard to spot a counterfeit in a large array of legitimate components. Therefore, it is critically important that customers fully and completely understand their supplier's business and technical practices prior to engaging in any use of the products. A capability model is particularly helpful in enforcing that understanding since it establishes a common and auditable basis between organizations.

The problems caused by breakdowns in the supply chain mirror the problems encountered in conventional manufacturing, in that the failure lies in the inability to do the work due to the lack of a component. The same is true with the technical service concern. From the standpoint of product security, a failure to deliver a critical part prevents the ICT product from being used, which is the equivalent of a denial of service in conventional security terms. Thus, efforts to mitigate security risks or risks to product integrity tend to concentrate on identifying and managing single points of failure. Capability models help in that respect because they establish common management functions designed to monitor and control the overall process of construction or maintenance.

From a technical service standpoint, the focus is on learning whether the supplier's operation is capable of delivering the product as specified. Since supplier capability is at the center of any acquisition or outsourcing decision, it is important to find out in advance whether the contractors that comprise the supply chain possess all of the capabilities required to do the work. Specifically, suppliers have to prove that they are capable of developing and integrating a secure product. Overall capability is usually demonstrated by the supplier's past history with similar projects as well as their documented ability to adopt good software engineering practices. A commonly accepted and fully auditable model of best practice shared by the customer and the supplier helps to cement that assurance.

The issue of unintentional vulnerabilities is just a specific application of the overall development and sustainment problem in that defects in software and hardware occur because of failure in the process. By definition, the installation of unintentional flaws is not a hostile act; however, since the problem is so pervasive, the sheer number of exploitable vulnerabilities placed in ICT products makes unintentional flaws and defects a major concern.

There is an extensive body of knowledge in ICT product assurance; however, since the steps necessary to ensure product integrity have to be instituted, managed, and sustained in a logical way, best practices are often not followed or performed half-heartedly. The result is that common defects in ICT products are exploited by a growing array of criminal and other bad actors. The installation and sustainment of a commonly accepted capability model addresses this concern directly. Nevertheless, it is critical that the activities in that model be executed in a continuous and disciplined fashion.

Putting Product Assurance into Practice

It should also be clear from the GAO report that the assurance of trustworthy products calls out three commonsense principles: control the development and sustainment work using common best practice, adopt rigorous assurance practices at the component level, and rationally plan for failure (Figure 1.3).

A very large percentage of the counterfeiting, supply chain critical point of failure breakdown, and capability concerns can be mitigated by simply ensuring that every one of the entities up and down the supply chain is under strict management control. Unwanted functionality and development failures must be addressed by strict product assurance from the time of inception to the time of acceptance. Then, when the inevitable failure does occur, there is a well-defined strategy in place to ensure that the problem is properly addressed.

An authoritative, mutually agreed-upon basis for the rational management of the project has to be available in order to independently assure organizational competence, while building commensurate trust. This requirement has two practical conditions associated with it. First, the supplier must be demonstrably competent, and second, the buyer must be able to reliably identify a competent supplier. If an acquirer has a long-standing history with a given supplier, that organization will know whether that supplier is competent. However, given distances and the global elements of business, this is not a common situation. As such, a defined process or standard is implicitly necessary to assess and certify competence, and the national standard for doing that will be presented in this book.

The common requirement for properly executing this standard is that the trust element be consistently enforced by some sort of universally recognized common management assessment or audit function. A standard and audited assurance of the

Figure 1.3 Three commonsense principles.

proper execution of a set of formal practices underwrites two of the most important factors in global business, namely, trust and competence. More specifically, a tangible and effective SCRM system can enhance the level of trust that exists between a supplier and a customer as well as provide the customer with documented assessment assurance that the supplier is competent.

Obviously, a big part of ensuring trust relies on the ability of the supplier to guarantee that it can deliver on cost, timeline, and integrity commitments. The issue is that given the complexity of most ICT projects, it is difficult for any supplier to provide that sort of guarantee. According to Watts Humphrey (1989), the three variables that serve as the basis for trust in business are history, understanding, and awareness. All ICT organizations have difficulty assuring any of these three factors because companies will normally deal with each other electronically over continents and oceans. Therefore, unless the acquirer and supplier have done business with each other before, those companies often have little basis for gauging performance. Additionally, even if two companies have experience with each other's work, there is no guarantee that a customer can rely on similar results in a succeeding project, given all of the factors involved in technology product manufacture.

The need for a single, fully defined, standards-based process infrastructure is a precondition for any subsequent execution of the SCRM process. The role of ICT management is to ensure that supply chain violations do not occur in the first place. However, since work is outsourced, the acquirer does not actually have hands-on access to it. Therefore, acquirers have to utilize some form of generic assessment-based management process in order to ensure product integrity. The aim of a management process is to establish a systematic approach to the way the organization goes about its ICT work.

An orderly process is important because it will ensure corrective action is built into the product development in the first place rather than tested at the end. Therefore, the insistence that an ICT management system conform to a commonly accepted model of best practice ensures best-of-breed management. Most leading-edge corporations have had such a management system in place for years. Now, with the current set of generic ICT best practice standards, any organization can implement the successful practices that more advanced corporations employ.

The Supply Chain and the Weakest Link

The software and services supply chains that underwrite our way of life constitute a significant avenue of attack. This is because the supply chains are probably insecure and nobody really knows for sure how secure any sourced product derived from a supply chain is (GAO, 2015). This fact alone is the justification for a comprehensive ICT SCRM practice and the mechanism for deploying ICT SCRM is built around enterprise-level strategic planning. ICT SCRM strategy ensures the integrity of complex product and service supply networks and ensures that all risks and single points of failure are mitigated for each supplier network to a sufficient level of satisfaction for all stakeholders present in a sourced-product supply chain. This book incorporates recommendations from four international standards:

1. ISO/IEC 12207-2008, *Software Life Cycle Processes*
2. ISO/IEC 27001:2013, *Information Technology-Security Techniques-Information Security Management Systems (ISMS)-Requirements* & 27002, *Code of Practice for Information Security Management*
3. ISO/IEC 31000:2009, *Risk Management-Principles and Guidelines*
4. NISTIR 7622, *Notional Supply Chain Risk Management Practices for Federal Information Systems* (NIST, 2012), NIST SP 800-37, Rev.1, *Guide for Applying the Risk Management Framework to Federal Information Systems: A Security Life Cycle Approach* (NIST, 2014) & NIST Special Publication 800-161, *Supply Chain Risk Management Practices for Federal Information Systems and Organizations* (NIST, 2015)

We will present the knowledge necessary to deploy a complete SCRM system (SCRMS) with the aim to outline both the context and the detailed practices that underlie that context. Starting from the conceptual framework, we will move down to the requisite activities and tasks. At the end of this book, the reader will understand how to

∎ Implement a comprehensive, well-defined, and organization-wide standards-based SCRMS
∎ Customize an appropriate set of SCRM activities for a given organization or project at the requisite level of process capability
∎ Organize, implement, and manage effective SCRM operations for a complex supply chain

Visibility and Control

We build systems out of components, which are derived from global sources outside of our direct control. Thus, a nation-state, terrorist group, or even individual who wants to compromise a supposedly secure system can easily and surreptitiously succeed through a third- or fourth-tier supplier. Thus, nobody knows for sure whether the parts that comprise our national infrastructure are actually what they were intended to be or whether they are counterfeit and possibly contain maliciously inserted objects (GAO, 2012).

ICT products are developed through a global supply chain with the purpose of supplying a product or service through coordinated work involving several organizations. The problem is that ICT supply chains produce products that are either abstract like software or so infinitesimally complex that they cannot be overseen and controlled by conventional means. Thus, ICT supply chains create a different set of assurance problems for managers. Proper ICT SCRM practices address those assurance problems by providing a consistent, disciplined environment for (Figure 1.4)

- Developing the product
- Assessing what could go wrong in the process (i.e., assessing risks)
- Determining which risks to address (i.e., setting mitigation priorities)
- Implementing actions to address high-priority risks and bringing those risks within tolerance

Typically, supply chains are hierarchical, with the primary supplier forming the root of a number of levels of parent-child relationships. From an assurance standpoint, this implies that every individual product for each individual node in that hierarchy has to be secure as well as correctly integrated with all other components up and down the production ladder. Because the product development process is distributed across a supply chain, maintaining the integrity of the products moving within that process is the critical part. The weak link analogy is obvious here.

Consequently, the activities within that product's supply chain have to be consistently rational and precisely controlled in order to ensure against sabotage or unintentional harm. This requires a coordinated set of consistently executed activities to enforce visibility in the process. In this respect, the purpose of the ICT SCRM function is to ensure the integrity of disparate objects as they move from lower-level construction up to higher-level integration.

Figure 1.4   Proper ICT SCRM practice.

Building Visibility into the Acquisition Process

Globalization is good for business, and it allows for a more competitive bidding practice. It gives companies access to the best talent available worldwide, but it also brings challenges. Because of the international nature of supply chains, organizations can interact by proxy with suppliers which they may not know about and/or may never see. And less insight into suppliers' security practices means less control over their business practices, which can mean increased vulnerability to adversaries. Moreover, supply chains must always satisfy stakeholder criteria.

Not only do we have an increasingly globally interdependent supply chain, the complexity of modern ICT products demands a myriad of capabilities. This creates a situation in which attackers from a nation-state, terrorists, criminals, or rogue developers might be able to gain control of systems through supply chain opportunities and intentionally implant logic or create unintentional vulnerabilities that could be maliciously exploited. The traditional consequences are loss of critical data, intellectual property, and/or technologies. The emerging consequences are potentially the intentional exploitation of manufacturing and supply chains that can result in corruption with the resulting loss of confidence in critical organizational capabilities, systems, and networks.

For key systems and networks, the aim is to manage risk to the critical components throughout the acquisition life cycle by installing proactive SCRM key practices to strengthen acquisition operations security, employ technical mitigations, enhance vulnerability detection, and partner with other organizations to drive security manufacturing, engineering, and test and evaluation practices. Mission-critical functions and components have to be integrated into their applicable system at a level of assurance consistent with the criticality of the system and the functions and components' roles within the system. Then the associated risk for each has to be identified and managed at a level of trust commensurate with the criticality of the component throughout the entire system life cycle.

Risk management is critical to the overall procurement process in that the processes, tools, and techniques deployed for any given organization must protect the quality, configuration, and security of the organization's systems, firmware, hardware, and systems throughout their life cycles. More importantly, the tailoring process has to include the assurance of components or subcomponents from secondary sources. The implication of this is a single coordinated process to detect vulnerabilities within custom-developed and standard commodity hardware and software by means of rigorous testing and evaluation activities including developmental, acceptance, and operational testing.

The assurance cannot just stop at the door. Because products are placed into day-to-day operation, there must also be a means to detect the occurrence of, reduce the likelihood of, and mitigate the consequences of unwittingly using products that might contain some form of flawed or malicious function or counterfeits. Therefore, a well-defined, designated element of system engineering that applies scientific and engineering principles to identifying security vulnerabilities and minimizing or containing risks associated with those vulnerabilities is required (Figure 1.5). This engineering process should be capable of

- Incorporating security requirements into the system engineering concepts and processes of the organization as well as integrating security requirements into all evolving system designs and baselines
- Identifying and implementing countermeasures and subcountermeasures to assess risks and determine mitigation approaches to minimize process vulnerabilities and design weaknesses
- Performing cost/benefit trade-offs to ensure affordability
- Ensuring that secure design considerations are an integral element of life cycle management of the ICT acquisition decisions throughout the full life cycle

Systems operate on a life cycle basis, and evaluation activities like criticality analyses need to be scheduled according to life cycle phase. This includes taking steps to evaluate overall business goals and purposes and identifying functions within the system architectures that constitute mission-critical elements. This also implies the need to maintain and refine a product tree comprising critical candidate hardware, software, and firmware subcomponents in order to ensure that all entities responsible for those functions are fully identified. The product tree of critical system components and subcomponents then has to be maintained and refined over time. This involves reviewing the list of critical system components and subcomponents to confirm that appropriate coverage is maintained in the configuration audits and the life cycle sustainment plan. The aim is to continuously ensure life cycle integrity.

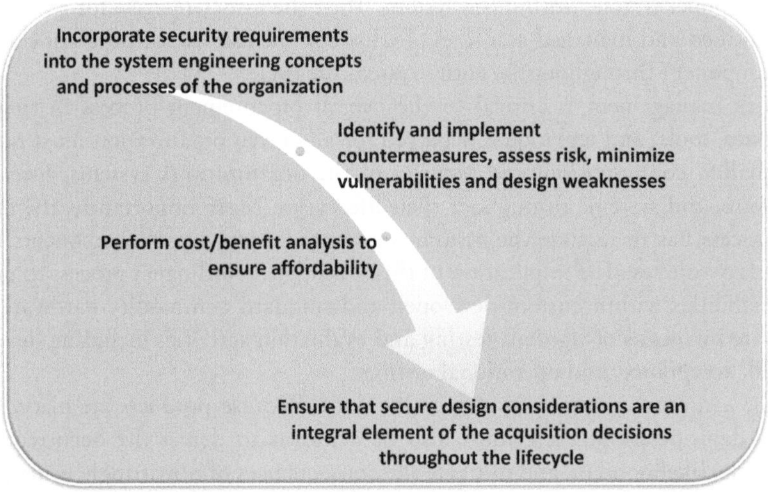

Incorporate security requirements
into the system engineering concepts
and processes of the organization

Identify and implement
countermeasures, assess risk, minimize
vulnerabilities and design weaknesses

Perform cost/benefit analysis to
ensure affordability

Ensure that secure design considerations are an
integral element of the acquisition decisions
throughout the lifecycle

Figure 1.5 Engineering process.

The Seven Phases of ICT Acquisition Practice

The acquisition of ICT is a strategic business process, not an isolated managerial activity, and is built around a logical sequence of seven managerial phases. The relevant activities in these phases need to be practiced continuously throughout the sourcing and procurement function. The process is also dynamic in that it adjusts its component actions and tasks to changes in the technology or risk picture as those occur in the day-to-day operation of the business. Naturally, the primary cause of that adjustment will be the appearance of new technology or threats.

ICT acquisitions are planned and deployed by means of a strategic management function, and the ICT acquisition operation is always maintained through a set of comprehensive and systematic policies designed to ensure optimum visibility and control. The ICT acquisition process is meant to be carried out as an organizational control process that is performed on a no-less rigorous basis than corporate financial control. That means that all relevant acquisition process documentation is maintained in auditable condition.

The standard ICT acquisition process has a life cycle that embodies seven fundamental stages or areas of practice (Figure 1.6):

1. Procurement Program Initiation and Planning
2. Product Requirements Communication and Bidding
3. Source Selection and Contracting
4. Supplier Considerations
5. Agreement Monitoring
6. Product Acceptance
7. Project Closure

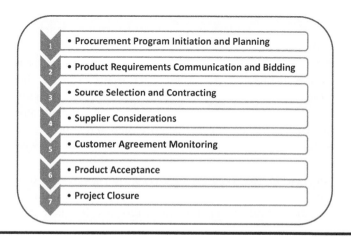

1. • Procurement Program Initiation and Planning
2. • Product Requirements Communication and Bidding
3. • Source Selection and Contracting
4. • Supplier Considerations
5. • Customer Agreement Monitoring
6. • Product Acceptance
7. • Project Closure

Figure 1.6   Seven phases of the ICT acquisition process.

Practice Area One: Procurement Program Initiation and Planning

Logically, the first stage in the life cycle involves acquisition program initiation and planning and is the point in the process when the actual decision is made to launch the acquisition process. Since the decision to acquire involves justifying a particular purchase decision, the first formal activity in the acquisition process entails identifying and documenting the need for the system, software, or even service product. This includes describing the business purpose(s) of the system and important business contextual factors such as cost benefit and risk. Any major acquisition will involve a huge commitment of time and organizational resources; therefore, from a business standpoint, the initiation and planning stage might be the most critical point in the life cycle. At this stage in the life cycle, the practical business advantage and risks of a future acquisition are evaluated and traded off against the need for a new system or service.

The environment and uses of a product will explicitly shape the form of the purchase. For example, the process for buying a piece of a national security system is going to look entirely different from the purchase of a game program for your home computer. Yet, both of these items are fundamentally just computer code. The source of that difference is that the contextual factors and the context substantively affect the decision-making. Therefore, this stage is driven by a prepared statement of the costs and benefits of the proposed acquisition, which is then developed and approved from a single coordinating control point in the organization.

There is a strong rationale for initiating all acquisitions through a single control point. This is because ICT system purchases can originate from many sources within the organization, and those sources can reside at many levels within the organization. Without a centralized vetting and approval process, it is difficult to ensure that every request for a new system or service is reasonably risk free, cost justified, or even necessary. Additionally, environmental factors often trade off against each other so all of the elements of the business, technical, and security justification have to be described to include the logic that went into the final documented decision. This includes documenting all feasibility and security issues. It should be kept in mind that the requirement to document a purchase decision applies to more than just new development. It should also apply to the decisions about any significant enhancements. For this reason, the first practical step in the software procurement process is for a central analysis function to prepare a fully documented general justification for every procurement.

The documentation itself typically involves a formal statement of the business advantage that will result from acquiring the product or service. More importantly for our purposes here, the documentation must also provide a thorough, contextual threat assessment and risk analysis. In essence, the documentation must provide a rational business and assurance case for the purchase, and it can serve as the context for managerial decision-making regarding whether to develop the product or purchase it off the shelf. Purchasing a product is an attractive option because COTS software represents a considerable business advantage over custom developing a product in house. Because the R&D costs are spread across the customer base, COTS products are generally less costly and are usually immediately available. However, all purchasers should understand the security problems with COTS products.

As we said earlier though, the primary problem with purchasing is lack of visibility into the product itself. Therefore, a well-defined and formally instituted risk identification and threat assessment is an essential part of the assurance case for any proposed COTS purchase. Threat assessment and risk analysis helps to ensure that the issues associated with acquiring a product are identified and criteria for selection that will satisfy all of the basic requirements for product integrity and security are defined. The consideration of risk in the project initiation stage is a key factor in developing the specific actions needed to ensure the security and integrity of the product from the start. From a business benefit standpoint, the identification and analysis of potential threat helps the purchaser make better decisions about the level of investment needed to address known risk.

The next step in the process is the preparation of the acquisition project plan. The chief advantage of having in place formal documentation of the costs, benefits, threats, and risks is that it supports the intelligent preparation of a comprehensive acquisition project plan. Having a complete roadmap of threats and risks prior to the planning stage allows planners to make intelligent decisions as opposed to simply launching into the unknowns of the marketplace without a strategy.

Thus, planning should take place only after the context surrounding planning decisions is fully understood. The context allows planners to accurately describe the project boundaries and constraints as well as the specific features and functions the product will provide. Statements of desired functionality are not the same as specifications of requirements and are more much general. They are often stated in the form of business requirements.

Although they are general in focus, these statements can be extremely valuable in guiding the acquisition process because there is always some uncertainty in purchasing products because the purchasing organization often does not know exactly what existing product features are available in the marketplace. The purchaser will always have to deal with the problem of buying what's available. Without a thorough market analysis, it will be difficult to develop a realistic statement of product requirements if it has already been determined that the new product will be a COTS purchase.

Unfortunately, planning without understanding risks creates a major security problem. Normally, once an organization makes a commitment to buy, it will buy something. Setting out to make that purchase with only a very vague idea of the threats and risks in the product domain makes it hard to select the right product. The tendency for management to leap before it looks raises serious long-term product-security concerns; this is the chief justification for documenting all of the issues associated with threat and risk as early as possible in the overall acquisition process.

Practice Area Two: Product Requirements Communication and Bidding

Once all of the factors required to align the acquired product with its intended purpose have been considered, the next step involves the definition of the actual form of the product, the formal requirements. The organization develops and documents an explicit set of functional product requirements in order to accomplish this step. The product requirements must fully describe in detail the actual shape of the product.

This standard document is normally attached to a formal request for proposal (RFP) document; it typically specifies the system requirements, including desired behaviors. The requirements specification expresses in clear behavioral or functional terms each explicit operation the system will perform. The process that underlies the development of this deliverable entails an operational and threat analysis that determines all aspects of the desired product behavior to include business and organizational risk factors.

In the latter respect, the System Requirements Specification (SRS) specifies all relevant safety, security, and other types of protection concerns including all relevant design, testing, and compliance factors. Thus, the formal specification of system requirements expresses all contextual factors, including business processes. The system context is important because it establishes the subsequent rigor of the confidentiality, integrity, and availability requirements of the system.

Practice Area Three: Source Selection and Contracting

The definition of system requirements is followed by the issuance of a set of written requests to prospective suppliers. The request documents the desired product characteristics as well as the terms and conditions for delivery, setup, and long-term support. The document is typically called an RFP, and it cannot be stressed enough that the RFP is a formal business artifact. The aim of the RFP is to provide a clear and unambiguous description of the product to all prospective suppliers. The artifact that communicates that description is the written and highly detailed SRS, which was prepared in the prior area of practice. The SRS spells out in legally enforceable terms the precise set of functional and security requirements that will be required to achieve a satisfactory solution.

The practical goal of RFP preparation is to allow everybody who is going to be involved with the system to contribute to its definition. Participation ensures both organizational and user buy-in when the system is delivered and installed. A satisfactory RFP will wholly describe two critical elements. The first comprises the system and security functions that must be delivered in the product. The second is the criteria that will be used to evaluate whether these functions have actually been delivered. In effect, the statement of functions required characterizes the externally observable behaviors on which the product will be evaluated at the time of acceptance. The criteria for evaluation itemize how the customer organization will utilize the product, to confirm that the desired functionality is present and correct.

All system specifications involve trade-offs. An acceptable final product will embody a competing set of mutually exclusive needs, such as cost, versus the number of functions. The competition has to be taken into consideration when the criteria for evaluation are developed. Risks have to be evaluated in light of the business environment in order to underwrite the decision-making with respect to the security requirements and the concomitant priority of the security requirements. Making the right decision about where to invest in security involves a number of related business factors. These include the technical soundness of the solution, its contribution to the business operation, cost/benefit priorities, and the overall resource constraint of the organization. Because there are a variety of issues, the ability to negotiate trade-offs is important.

Consequently, the final contents of an RFP will include the most acceptable specification of the system and software requirements for the organization (Figure 1.7):

- The assurance requirements to be included in the eventual statement of work (SOW)
- Any inherent certification and accreditation requirements

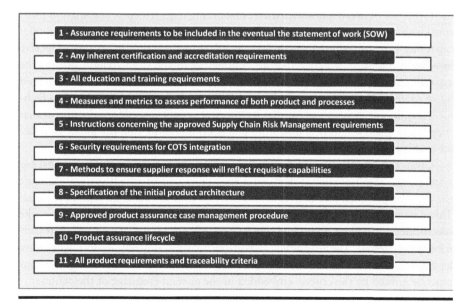

Figure 1.7   Contents of an RFP.

- All education and training requirements
- Measures and metrics utilized to assess performance of the product and process
- Instructions concerning the approved SCRM requirements
- Security requirements for COTS integration
- Methods to ensure that the supplier response will reflect requisite capabilities
- A specification of the initial product architecture
- The approved product assurance case management procedure
- The product assurance life cycle
- All product requirements and traceability criteria

The aim is to create a level playing field. The RFP ensures full understanding of the standard product functions and assurance criteria to be embedded in a legally binding contract. The type of contract created depends on the criticality of the system as well as its relative size and complexity. The contract always comprises a trade-off among the level of acceptable cost, schedule, and performance, in conjunction with the risks that are acceptable for both the acquirer and the primary contractor. The contract language explicitly spells out the complete set of assurance requirements including how each requirement will be tested and evaluated to ensure final acceptance. More importantly for our purposes here, the contract is where all outsourcing considerations are spelled out and conditions for proper execution are specified.

Outsourcing represents one of the most critical areas of risk in the acquisition process because the contract provides the only legally enforceable means for assuring supplier accountability. Therefore, it is especially important to understand and then spell out in detail any anticipated outsourcing requirements. The analysis must consider a number of standard risk factors. First, as the GAO has stated, supplier capability is at the center of any outsourcing decision; it is thus important to receive in advance tangible evidence that all contractors in the supply chain possess the capabilities to do the work. An acceptable level of capability is usually documented by an examination of the supplier's past history with similar projects. From a technical standpoint, it is important that prospective suppliers be able to document that they follow good software engineering practices, including that the prime contractor is able to assure the work of any subcontractors he or she might employ.

Given the complexity and the analogous requirement for capability of most big systems, the acquirer should demand a specific record of which system elements the supplier prepared and delivered versus those that the subcontractors provided. This statement is necessary to allow the acquirer and the prime contractor to make specific determinations about how any subcontractors were managed and their contributions assessed. The assignment of work elements to subcontractors must always be explicitly stipulated in the contract, and all additional outsourcing decisions must be guided by the criteria embedded in the general language of the contract.

In addition to outsourcing, a strategic concern has to be addressed in the development of the contract—the determination of the degree of foreign influence and control that might be exercised over the product. Influence and control by organizations or nation-states that are not necessarily friendly to the US will directly impact the trustworthiness of any product. Therefore, the degree of foreign influence, control, or ownership of a given contracting organization has to be determined and ensured before additional steps can be taken to manage risk.

The specific aspects of risk have to be investigated once the business context is fully understood. It goes without saying that a robust risk management approach is the norm in all modern contracts. Therefore, the contract has to provide details of the risk management process. Those details include the specific risk assessment mechanism, the method for designing and deploying risk mitigations, risk mitigation monitoring activities, and how the risk mitigation activities will be adjusted to meet changing conditions. Plans to address these factors are normally included in a contract (Figure 1.8):

- The specification of the detailed life cycle model
- The specification of processes, activities, and tasks as mapped to the life cycle model
- The specific approach to ensuring the quality and security of the project
- Product assurance criteria
- Scheduled product assurance reviews and audits
- The supplier's specific method and approach to product assurance

Factors to Include in a Contract
1. Specification of a detailed lifecycle model
2. Specification of processes, activities, and tasks mapped to the lifecycle model
3. Specific approach to ensure the quality and security of the project
4. Product assurance criteria
5. Scheduled product assurance reviews and audits
6. Supplier's specific method and approach to product assurance
7. Milestones for assurance reviews
8. Performance criteria if any form of a service level agreement (SLA) is utilized
9. Any product certification and accreditation requirements
10. Specific management means for controlling product architecture evolution
11. Frequency of risk assessments and risk management plan updated
12. Frequency of product assurance risk evaluation
13. Specific escalation mechanism for elevating product assurance issues
14. Specific issue resolution plan and processes
15. How corrective actions will be monitored
16. Capability requirements for product assurance professional staff
17. Identification of key product personnel
18. How key personnel will be monitored
19. The monitoring process for the assurance training program
20. How staff experience level will be monitored

Figure 1.8 Elements to include in a contract.

- Milestones for assurance reviews
- Performance criteria, if any form of service level agreement (SLA) is utilized
- Any product-explicit certification & accreditation (C&A) requirements
- The specific management means for controlling product architecture evolution
- How often risk assessments will be done and the risk management plan updated
- How often product assurance risks will be evaluated
- The specific escalation mechanism for elevating product assurance issues
- The specific issues resolution plan and process
- How corrective actions will be monitored
- Capability requirements for product assurance professional staff
- How to identify key product personnel
- How key personnel will be monitored
- The monitoring process for the assurance training program
- How staff experience level will be monitored

Practice Area Four: Supplier Considerations

Practice area four shifts the focus to the supplier up to the point the responsibility for ensuring proper practice rests with the acquirer. Once the contract is signed, the project is in effect transferred to the supplier. There are several options at this point, depending on whether the product is purchased, whole or in part, off the shelf, or whether the product is to be developed as part of a project. If a service is the outcome of the contract, then the responsibility for the service-level assurance has to be agreed on and the responsibility assigned.

If the product is to be developed via a supply chain, the first step involves the development and documentation of a project management plan. The supplier is responsible for the implementation and execution of that plan, and therefore it is customary for that entity to draw it up in detail. Although this is most commonly done in conjunction with the acquirer's management team, at a minimum, the acquirer is responsible for the definition of a set of monitoring and control functions that will be performed throughout the contracted life cycle.

At this point in the process, the supplier in conjunction with the acquirer will make the decision whether to develop the product using the supplier's internal resources or to develop the product by subcontracting. Obviously, the essence of good subcontractor management rests on the correct definition of the supply chain process, the precise identification of the participants up and down the chain, and the assignment of responsibilities for assurance and control. The most important aspect of the process is the decision about which off-the-shelf software products might be purchased and integrated into the final product from internal or external sources.

The next step is a complete and detailed definition of the monitoring process if the product is produced via a supply chain or the vetting of the suppliers in that chain if the product is already built. If monitoring is required, then there has to be a functional and clearly understood process developed to manage and control the subcontractors. This process has to ensure that all contractual requirements are properly and correctly expressed and then expeditiously passed to the subcontractors up and down the chain. Additionally, there has to be a clear-cut mechanism in place to ensure that the project interfaces with any independent verification, validation, or test agent that might be utilized to provide third-party assurance of supplier capability and practice. The aim is to make the contract detailed to the extent that the interface with all of the parties involved in making the product, along with the lines of responsibility, is unambiguously defined in the supplier's project plans.

If the product is being built through a supply chain, there has to be a defined organizational presence in place to coordinate and enforce all contracted review activities across all identified interfaces up and down that chain. This has to be established as part of the product assurance process in the contract. The activity itself will normally involve some form of joint review between the supplier and its subcontractors and between the supplier and the customer. This is typically conducted in accordance with commonly accepted definitions of best practice as normally specified in the contract.

The exact same requirements apply to a product that is delivered by a supplier as built, e.g., COTS. Some type of formal verification and validation activity has to have been specified and carried out in the construction process in order to ensure that contract requirements for both product assurance and process correctness have been met. The specifics of that activity are normally described in the contract. Nonetheless, there has to be an entity responsible for ensuring that those tasks have been performed and if there has been any deviation from contract specifications to enforce any corrections or variances required.

The final aspect is planning the shape of the reporting function. All reports have to be available to all parties as specified in the contract. These reports constitute the project documentation, and they also serve as the basis for assurance of product security, both collectively and individually. They provide the necessary assurance to top management that their acquisitions are properly overseen and correctly integrated into the overall organizational system. Finally, the product documentation is the basis for all corrections and lessons learned that might be required in product assurance over the long term. The assurance essentially makes the case that the supplier has provided all of the requisite parts and services needed to satisfy the contract and that the product itself can be trusted.

Practice Area Five: Customer Agreement Monitoring

There is a set of standard activities that the supplier will undertake, both prospectively and retrospectively, to ensure the integrity and correctness of the product construction process. The particulars are normally specified in the contract: standard considerations such as timeline, acceptance criteria, problem resolution processes, escalation procedures, and reporting lines. In addition, there has to be a set of routine assurance activities in place that will allow the acquirer to monitor the supplier's overall execution of the process and the general set of activities that are associated with assurance.

This is normally based on and defined by the formal system assurance process activities specified in such standards as ISO 12207-2008 and ISO 25010-2011. The general aim of this element of assurance is to supplement the regular contractually defined monitoring activities with specially targeted verification and validation actions that apply to the acquisition project as needed, no matter whether it is a development or a COTS acquisition. The aim of this part of the process is to make certain that all of the necessary management and technical information to guide project decision-making is provided in a timely and useful manner.

Practice Area Six: Product Acceptance

Whether the contracted product is a development or delivered as built, it is necessary for the acquirer to have already prepared an acceptance process. This process is always based on the acceptance strategy outlined in the contract. The acceptance strategy cannot be vague or imprecisely stated since there are reputational and fiscal implications for every product; therefore, there can be no doubt about what has to happen in order for the acceptance to take place.

Normally, explicit test cases and test data that allow for the emulation of an acceptable level of performance have to be specified. Since the acquirer knows what it wants in terms of that performance, the basis for carrying out the acceptance testing process is always laid out in the contract. This includes a detailed statement of the test procedures that will be conducted and the specific contextual details of the testing environment, both bench and operational. Since this is an acceptance function rather than one associated with development/delivery, the contract also needs to specify the extent of supplier involvement in the acceptance work.

There is normally a defined period of time during which the acquirer will carry out the intensive acceptance review and testing of the deliverable. Then, once all of the conditions for acceptance as stated in the contract are satisfied, the acquirer will accept the product from the supplier. From an assurance standpoint, it is necessary to accept the product both in terms of its ability to satisfy functional performance criteria and contractually specified security criteria.

Once the product is put into operation, it is necessary to develop and execute a secure sustainment plan. In some respects, the plan is as important as the actual product acceptance testing because it is normal for a product to develop vulnerabilities because of uncontrolled patching. Therefore, a formal configuration management process needs to be present. The customer must be made responsible for developing and maintaining strict configuration control, and the product should be seamlessly transitioned from the supplier's configuration control system over to that of the acquirer.

Practice Area Seven: Project Closure

Project closure activity is as important in some respects to the long-term viability of the system as the initial project setup process. Based on the results of the acceptance phase, the customer will make payment or provide other agreed-upon consideration to the supplier in accordance with the terms and conditions of the contract. The closure activities also require the supplier to "fit" the delivered system into the current operation of the customer. This is normally a straight-up system engineering activity, but it will also include all necessary training and awareness.

This last phase is often passed over, but it is critical to the security of the product because this is where the important good housekeeping and other operational security practices get embedded into the organization's operational culture. Therefore, the installation process is normally specified in detail in the contract along with all of the necessary deliverables.

Once the contractual agreement has been blessed by the acquirer's actual payment of funds owed to the supplier, there has to be some formal means of achieving mutual agreement that the project has ended and satisfactory payment made. This is a necessary final step because the responsibility for both the functional operation and the long-term security of the product has to be acknowledged as fully transferred to the customer domain. This is true for both products and services delivered. Of course, that does not mean that the customer and acquirer will sever all ties when it comes to the product; it just means that all forms of responsibility for the ongoing conduct of system functional and security operations will be left to the acquirer.

There is a tendency for actions that are necessary to ensure the long-term security and viability of the product to fall between the cracks with suppliers and acquirers. Therefore, the supporting actions and accountabilities following final payment and installation have to be explicitly spelled out in the contract.

Building the Foundation: The Role of Governance in Securing the ICT Supply Chain

ICT SCRM is founded on the creation of an explicit management control structure and a concomitant set of controls. A single coordinated framework of controls is necessary, because business has developed a growing dependence on COTS technology, and with that increasing reliance comes the increased need to ensure that the technology being purchased is properly secure. Thus, as business organizations grow and diversify, progressively more powerful means of ensuring trust up and down the supply chain are required.

In day-to-day application, this means that it must be possible for companies to assure every product that they source; this is a particularly difficult task to perform, given the layers of complexity involved in a global supply chain. Most product creation work is done in multilayered, multivendor, and even multicultural team settings, and all of the sourcing for the products integrated into a final system must be fully coordinated and controlled. Coordination and control of complex work necessitates a communal, shared point of reference that will allow the managers in the customer organization to benchmark the activities up and down the supply chain against best practice. This common point of reference is embodied in a security governance framework.

As such, one of the central elements of this book is the assumption that proper SCRM is founded on the principles and practices of strategic governance. The purpose of adopting this belief is to avoid a typical piecemeal solution. In essence, every organization operates based on a set of communally adopted practices, and it makes no difference whether those practices are even documented. If the organization does not adopt a single coordinated method for the creation of a universally standard set of practices, the performance of its operational functions will typically be based on some individual's or manager's understanding of the proper way to carry out that specific task. Those approaches tend to embed themselves in an organization over time, and their execution is more a matter of "company tradition" than of any particular form of logic.

Typically, the ICT SCRM activities that take place in any given corporation evolve in this fashion. The base practices develop bit by bit as the situation arises rather than being developed as an across-the-board planned system, with a well-defined purpose. The end result is that there is no actual assurance, or trust, because the way that the product is acquired does not fully embody or address all of the security concerns that might have existed in its creation and integration.

The alternative to a piecemeal approach is a formally defined and implemented infrastructure of comprehensive risk management practices, which are specifically aimed at optimizing the state of the security assurance and the level of trust within the supply chain. This is the primary argument for adopting a comprehensive governance and control-based solution to SCRM.

As with any complex deployment, the process of creating a complete governance framework is initiated through a rational and explicit planning process. The long-standing organizational rule that the approach to strategic governance has to be comprehensively planned also requires a formal statement of commitment and direct support from the top of the organization.

The prerequisite of strong executive sponsorship is critical to the success of the type of management we are talking about here. Such a strategic approach to the development of a security management infrastructure for a function as mundane as technical purchases will almost always represent a radical change in the way the organization views the acquisition process. Therefore, the natural human urge to resist changing a comfortable routine is frequently one of the most significant barriers to instituting anything as potentially far reaching as a comprehensive SCRM process. As a result, the simplest and most straightforward way to overcome middle-management resistance is to initiate the governance process from the top.

It is critical that the SCRM process is planned and executed as a strategic initiative. This is because technical sourcing is both organization-wide in application and has to be embedded in the operation as a long-term operational activity. So, we need to spend some time explaining what the general principles embodied in SCRM are and how they work.

The term *supply chain risk management* was coined to describe the strategic process that underwrites due diligence in the assurance of the organization's sourced ICT assets. Supply chain risk governance, or in common usage "management," deliberately builds a structure of rational, interorganizational relationships and common controls that can be used to manage and ensure trust in the sourced assets of any given organization. SCRM establishes a tangible mechanism to support a trust relationship between the company's suppliers and the purchased assets of the information-technology operation.

The means for assuring a proper level of trust is characterized by a relatively new term, "supply chain risk management." This approach centers on the creation of a comprehensive and persistent set of standard organizational controls, as well as a culture of assurance within the ICT purchasing process itself. This is opposed to an approach based on separate individualized assurance solutions for each unit and its products. Essentially, the business defines a coherent organization-wide framework of management functions, which embodies all of the necessary strategic policies, management roles, and organizational control behavior.

Five processes underwrite the development of a supply chain assurance solution to any given acquisition. The first of these processes is, "scoping." This term describes the intentionally planned and systematically executed process of establishing the boundaries and criteria of any sourced acquisition. In some respects, this is the most critical step since the depth and actual links in the supply chain will determine the form and extent of the rest of the solution. In the real world, that implies a conscious balancing act between the need for trust in a product and the resources that are committed to establishing it. Accordingly, the underlying issue the establishment of the scope of the supply chain will address is: "How does the organization get optimum assurance in the purchase of any given product?" In day-to-day practice, this means that it must be possible to make an intelligent decision about the level of risk that can be accepted for the sourced products available.

Anything can be assured if enough money is thrown at it; however, no organization has the money to effectively put a cop on every street corner when it comes to policing a retrospective supply chain. So, a deliberate process has to be undertaken that balances deployment of the controls to assure against supply chain risks and the likelihood and material consequences of the threat space. Factors that might enter into this process include considerations such as the level of criticality for each product and the degree of trust and concomitant assurance required for that product.

The determination is often captured on a 10-point asset classification rating scale, which can range from "who cares" on one end of the scale all the way up to "the business would close if this product proved untrustworthy" on the other end. Other issues might include the estimated impact of any prospective failure in the supply process as well as market, or regulatory, conditions as they impact product assurance. For instance, a product that has military applications would have a different standing on that scale than one that was purchased to support the operations of a hardware store.

The actual risk assessment activity itself is embodied under the next principle. However, this provides the opportunity to note that the process of purchasing any piece of technology is always dynamic. This affects the scope of the supply chain risk assurance needs because the product space for any given piece of technology is always subject to ongoing change and refinement. Assessment of the implications of that change is obtained from information fed back by ongoing, routine assurance activities, particularly threat and risk assessments. In essence, then, although the first step in the process is to define the scope of the SCRM solution, these rules and criteria are not fixed; they are always subject to change as the realities of the threat environment circumstance dictate.

The products that must be trusted and assured are defined by the particular protection issues associated with the product as well as the organization's resource constraints. In the worst-case scenario, that might involve purchasing a low-risk product without any consideration of trust or assurance. However, if that decision is made based on precise knowledge of the impact of the associated risk, it is an informed choice. In the hands of a capable decision-maker, the principle of scoping maximizes resource utilization and provides the foundation for the rest of the process.

The second SCRM process is *assessment*. This principle just signifies that the organization understands the nature of the threat environment, both internal and external. Because of the invisibility and complexity of ICT products, assessment is a primary player in any ICT-SCRM process. Assessment generally involves identifying gaps in product purchasing performance and then aligning current SCRM practices with some sort of best practice reference model, which is usually embodied by the expedient of a well-defined and commonly accepted standard. The substantive processes that are then deployed reflect strategic SCRM best-practice models such as ISO 28000, ISO 31000, and NIST-IR 800-161.

Depending on the outcome of the assessment, there might be a long period of trade-offs and refinement before an eventual resolution can be reached. However, the final documented process solution must always be capable of resolving the implicit vulnerabilities identified during the assessments. Accordingly, the end product of the assessment process always entails a concrete and coherent set of real-world processes and activities. The processes and their associated activities are installed and operated once the problem space has been sufficiently understood. The form of the actual recommendations for action might be nothing more than a simple statement of procedure, or they could be expressed using any of a number of formal specification, or even modeling, methodologies. The documentation that is the outcome of the assessment itself is typically a set of detailed descriptions of the substantive processes, activities, and tasks that will constitute the solution.

The *risk evaluation* then evaluates the impact of the identified threats on the proposed purchase as well as the organization's overall system assets. A particular threat may not necessarily have much negative impact on the system within a given organization's environment. So, the entire set of identified hazards is examined in order to differentiate those threats that would create specific and undesirable impacts. These are then carefully assessed with respect to their short- and long-term resource implications. Those assessment ratings are then recorded and prioritized so that the threats with the most critical impacts are dealt with first.

The means of dealing with threat is idiosyncratic, since threat itself is almost always unique. But the principle involved is twofold. First, all priority threats have to be countered by substantive and well-established operational controls. Second, there must be an organizational mechanism in place to keep those operational controls accurately aligned and functionally correct throughout the life cycle of the operation of each sourced technology. In order to keep that alignment, there has to be a persistent operational planning and evaluation mechanism to ensure the day-to-day supply chain risk process is functioning as intended. The mechanism is installed by a formal design function.

Therefore, the next principle is *design*. The outcome of the design process is a complete and proven effective day-to-day set of practices that address each issue identified in the risk assessment and fulfill all known constraints in concrete terms. Designs are not necessarily technical. In the case of a business model for ICT SCRM practice, the design might be a strategic plan or a policy and procedure manual. Nevertheless, the design process is always a creative, conceptual endeavor in the sense that its outcome is the complete and correct abstract model of the required security solution.

All designs exhibit common characteristics: they are complete when they encompass the entire solution and correct when all elements that should be present in the solution are provably there. They are understandable when how they communicate the form of the solution and all of the management control elements in the design is traceable to the issues raised in the risk assessment.

This is normally represented in a *design document*. The design document establishes the concretely assessable foundations for the measurement function. Additionally, the design also explicitly calls out the qualitative elements that will be employed to judge the success of the design itself. The specification that the purchased system must be proven "reliable" is an example of a qualitative element in that reliability can be defined as "up-time" percent.

In addition to direction, the design must provide a clear reflection of the anticipated degree of integration. Technical elements must be integrated at their interfaces to promote the most efficient interaction. Business elements have the same requirement at the process, activity, and task levels. All of the necessary relationship issues must be identified and resolved in the design. And as the name implies, along with the design there is always some sort of formal planning process involved.

Because a plan provides the explicit direction for the integration process, it is the essential end product here. For a small business, this might be a relatively trivial documentation item, a memo of agreement for instance. However, where the security solution is very large or complex, there is an implicit element of long-range planning and the result is always a detailed plan. In that case, the organization carries out the activities associated with a classic strategic planning process. This includes rationalizing the solution against business goals and long-term trends, as well as the formulation of a schedule and a contract. The outcome actively and intentionally aligns the form of the response to the organization's needs (Figure 1.9).

The largest principle, in terms of the actual time spent is *integration*. This concept might be just as appropriately termed "realization" or "implementation" because its product is the actual, substantive supply chain based product management process. For SCRM, the general goal is to ensure that the means employed to identify and provide the requisite processing and storage embodies the proper set of controls. This type of control-based assurance is typically embodied in the integration of five common attributes of security:

- *Authentication*: Whereby the risk control system has the capability to ensure that all products can effectively verify their identity when required.
- *Authorization*: Whereby the control system has the capability to ensure that all products are able to appropriately regulate access to a specific system resource once identity is properly established.
- *Confidentiality*: Whereby the control system has the capability to ensure that all products are able to maintain the secrecy of the contents of a transmission between authorized parties.

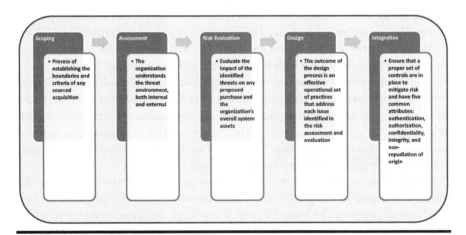

Figure 1.9 Five processes to develop supply chain assurance.

- *Integrity*: Whereby the control system has the capability to ensure that all products are able to assure that a transmission will arrive at its destination in exactly the same form as it was sent.
- *Nonrepudiation of origin*: Whereby the control system has the capability to ensure that all products are capable of ensuring that the origin of an authenticated electronic communication cannot be denied.

These common attributes are implemented and ensured by the overall SCRM control system. The general purpose of the SCRMS is to ensure against breaks in the links of the supply chain and to underwrite trust in all of the organization's sourced products. Accordingly, in most organizations, implementation of the SCRM process takes place in eight discrete stages. These stages are designed to force the organization to think through the details of process implementation, as well as to ensure that the solution is both optimally resource efficient and continuously correct.

The *first step* in the process logically involves the gathering of all of the information necessary to identify and categorize each of the supplier resources that will fall within the system. Essentially this involves the identification, labeling, and valuation of each and every one of the organization's ICT assets and the formulation of these into a comprehensive baseline inventory of products and their suppliers. In many respects, this part of the process resembles the organizational asset identification activity that would drive any fiscal accounting or physical inventory process, in the sense that individual categories of ICT product are simply differentiated and accounted for in a permanent ledger along with the sources of each product. Then this baseline ledger is maintained under the dictates of rigorous, management control.

Once all of the organization's information assets have been identified and arranged into a catalog of coherent baselines, the *next step* is to determine their current protection requirements. The purpose of this activity is to define the exact assurance needs for each of the individual items in the product/supplier baseline. This involves using the results of a risk assessment process to characterize all of the direct threats, vulnerabilities, and weaknesses for each baseline item. The outcome is that all of the factors that might impact the security of the product and its suppliers are understood and factored into the SCRM control system. When the entire set of product/supplier assets and related controls has been identified, the organization's decision-makers undertake a rational planning process to establish the criticality of each item. This step weighs the implication of all relevant risks that failure in the security of the product might represent, against the actions required to address them. In pragmatic terms, this means that the potential business impacts of the threat are evaluated against mitigation costs, and a level of significance is assigned based on the difference. This activity applies to any identified inherent product vulnerabilities as well as other substantive legal, contractual, and/or business issues. The working principle here is that the probability of the problem actually occurring must be balanced against the effort required to eliminate it.

The eventual outcome is a sliding-scale interpretation for each product/supplier in the baseline. This scale underwrites a risk reduction decision that might range between not dealing with the threat at all and deploying all necessary resources to completely eliminate it. This information is then aggregated into a total picture of the actions that must be taken in order to ensure the security of the sourced products in a baseline set of items. That then leads to a decision about the organizational resources that will be committed to implement assurance of product security for each individual product/supplier combination in the baseline, as well as the form of the overall control response itself.

One caveat has to be mentioned. This activity is not a simple matter of patching the product, since it is likely that the risk assessment will also identify threats up and down the supply chain that the organization would never have the resources to address. Instead, decision-makers perform a triage activity on each of the critical products in order to prioritize the entire set of security concerns for their supply chain. The goal of this activity is to ensure that the organization's critical trust requirements are adequately ensured within the resources that are presently available.

The *next three steps* are really aspects of the same process, which is the actual establishment of the day-to-day control system. Once all of the elements of supply chain risk and concomitant resource commitments are understood, it is time to create and array the initial set of controls for each product/supply chain the organization chooses to manage. This amounts to an iterative control validation and enhancement process. Over a period of time, the organization fine-tunes the controls that it believes are the most effective in ensuring product trust. Alternative approaches might be developed based on feedback from the stakeholders; normally this involves a significant period of time. Once the final set of controls is planned and proven correct, they are kept in a well-defined security baseline that is essentially standard operating procedure for SCRM.

The *final concept* in this process is measurement. In a technical setting, this might be called a metrics program. Because ICT and network products constitute a virtual asset, measurement is the fundamental element that enforces management control. Quantitative assessment of the security functioning of a given supply chain, in the form of such measurement items as breaches, breakdowns, or identified flaws and/or malware, helps decision-makers evaluate the ongoing performance as well as assign accountability. Therefore, every overall SCRM governance scheme must be accompanied by a description of the measurement program. Basically, this requirement just ensures that confidence in a given product can be ensured through measurement data. In that respect, then, measurement provides the consistent ability to confirm that an appropriate set of controls is in place and functioning properly.

Measurement programs are typically founded on a range of standardized metrics. These allow managers to evaluate performance at any given point in time in a given situation. Properly set up and maintained, the measurement program will monitor every aspect of the SCRM function and bring any deviations to management's attention. The actual data gathering is supported by regular tests and reviews of operational elements in the SCRM process done at preplanned and mutually agreed-upon points in time.

Thus, an explicit plan for conducting verification and validation activities and their associated targets is an essential component of any SCRM implementation plan. There are no universally recognized standard metrics for assessing the performance of a supply chain and assigning trust. Instead, individual organizations decide on and deploy measures they feel best fit their particular situation. The rule for this is straightforward; whatever metrics are selected must be uniformly and consistently applied and NOT subject to misinterpretation. In particular, since supply chain security operations are oriented toward product assurance in a network of suppliers, there is a need for a uniform definition of what constitutes an objectionable result. This requires the organization to delineate a measurable characteristic of actions that it considers to be improper activity. This statement is not a black-and-white proposition, so the actions that a particular organization considers to be improper need to be expressly clarified. That clarification has to be repeated for every potential security concern in the supply chain in order to make the data produced in the assessment accurate and meaningful. This clarification is especially necessary in order to help the measurement program function properly. It simply requires that the organization prepare and disseminate a standard classification and definition of all undesirable or improper actions in the product-sourcing process. Moreover, during the process of thinking these actions through, the proper individual metric items can be identified. With that approach, it is possible to achieve a consistent, measurement-based description of the security function.

The organization can engineer, or at least think through, its SCRM solution by following the activities that have been outlined above. The end result of this process will be a coherent and fully integrated security response. Nevertheless, this is never a one-size-fits-all process. Instead, the determination of each of the activities and tasks in the individual SCRM process must be driven by assessment and managerial decision-making about the degree of assurance required for a particular product within a given setting (Figure 1.10).

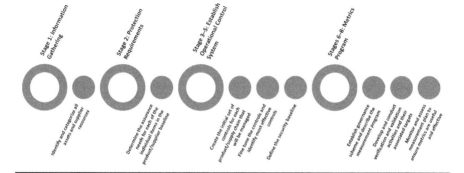

Figure 1.10   Implementation stages of the SCRM process.

The Use of Standard Models of Best Practice

The rule that underlies that condition is that all situations are different; thus, a correctly developed SCRM infrastructure is always precisely tailored to correspond to the exact requirements of the circumstance. In practice, there are governance models appropriate to secure different levels of the operation, and they are meaningful to differing groups within it. The end result of the diverse range of guidance is that the effective purpose and precise application of each model has to be understood in detail to be used properly.

The recommendations embodied in a best practice standard are meant to provide a concrete basis for characterizing a wide range of abstract assurance functions, including intangibles such as product information flows. In concept, the recommendations of a best practice standard are not standalone elements. They actually substantiate numerous facets of the day-to-day assurance operation that taken together constitute aggregate best practice. Accordingly, as a set they comprise a complete and tightly integrated organizational system that when properly configured and deployed produces a fully trustworthy set of sourced products. In that respect, then, best practice standards provide a formal mechanism that allows the organization to specify and associate the rules for control as well as describe how ongoing status will be assessed.

There are nine commonly recognized factors, which taken together are considered to be instrumental to the successful creation of an organization's SCRM assurance system (Figure 1.11). They are as follows:

1. The existence of a formally documented security policy
2. An implementation model that fits with organizational culture
3. Visible support and commitment from management
4. Organization-wide understanding of security requirements
5. Formal risk assessment and risk management procedures
6. Effective marketing of security to all managers and employees
7. Appropriate training and education
8. A comprehensive system of measurement
9. The provision for feedback and suggestions for improvement

This suggests that an organization must originate the policies to fulfill its aims and purposes within its particular culture. This is embodied through a governance approach that centers on a coherent framework of selected controls. This governance process requires that there be some kind of standard point of reference to coordinate all security activities that will be carried out up and down the supply chain. This is the reason every one of these approaches requires a standard model of expert best practice. The rest of this book will be devoted to that discussion.

1. A formally documented security policy

2. Implementation model that fits organizational culture

3. Visible support and commitment from management

4. Origination-wide understanding of security requirements

5. Formal risk assessment and management procedures

6. Effective marketing of security to all stakeholders

7. Appropriate training and education

8. Comprehensive system of measurement

9. Provision for feedback and suggestions for improvement

Figure 1.11 Nine factors for a successful creation of an SCRM assurance program.

Chapter Summary

The goal of this chapter is to demonstrate how a formal approach to acquisition and supply chain security can be used to assure the integrity of the technology base of any organization. The key concept here is "across-the-board trust." Because *all* of the potential components of the technology base are involved in the secure functioning of the system, every aspect of that base must be dependably secure, or "the weakest link" applies. The reader will discover how formal processes and a standard baseline reference are necessary to establish that requisite level of trust.

An organization's ICT procurement function is no different from any other purchasing function, in that the aim of any procurement activity is to acquire an effective product for the organization. Consequently, whether the product is a video game or a piece of sophisticated military hardware, the activities that take place within the acquisition process have to be logically related, controlled, and coordinated. A standard model of the ideal practices that are carried out by that process simply ensures that the control is implemented systematically and is effectively maintained through the specific actions of the individuals who are responsible for performing the assigned task.

From a security and integrity standpoint, this implies that every individual action in the overall process has to be rationally and properly placed in the timeline for execution. In addition, each task must be fully and correctly integrated into the overall activity. Therefore, at its core, the acquisition process we will talk about must be well defined and properly executed. It must ensure that the proper relationships are maintained among the larger set of actions that has been arrayed to achieve a given purpose.

The ideas presented in this book are well-established elements of a single process that has been developed by the federal government to ensure trusted product acquisition in its space. Specifically, the book presents the concepts of ICT SCRM from the perspective of NIST IR 800-161, the first standard body of knowledge for secure SCRM.

The GAO summarized the security concerns associated with ICT organizations in a March 23, 2012, report. ICT issues fall into five categories. Each category has slightly different implications for product integrity. "Installation of malicious logic on hardware or software, installation of counterfeit hardware or software, failure or disruption in the production or distribution of a critical product or service, reliance upon a malicious or unqualified service provider for the performance of a technical service and installation of unintentional vulnerabilities on software or hardware" (GAO, 2012, p. 1). Malicious logic is embedded in a product to fulfill some specific purpose. Malicious objects are by definition not part of the intended functionality. Therefore, in order to find and eliminate any instance, rigorous testing and inspection are required. Embedding a malicious object in a product is always a hostile act. Thus, the assurance that a product is free of malicious code should be a high priority with any ICT customer.

Counterfeits are not just an acquisition issue. Counterfeit parts can appear at any stage in the development and sustainment of ICT products. Counterfeits execute product functions as intended. They threaten product security and integrity because they are not the same as the actual part. Generally, the purpose of a counterfeit is to save money or supply a feature the maker is otherwise incapable of providing. As a result, counterfeits embody shortcuts in product quality or security that can fail in many ways. Because they function like the original part, it is often hard to spot a counterfeit in a large array of legitimate components.

The problems caused by breakdowns in the supply chain mirror the problems encountered in conventional manufacturing: the failure lies in the inability to do the work due to the lack of a component. The same is true with the technical-service concern. From a technical-service standpoint, the focus is on learning whether the supplier's operation is capable of delivering the product as specified. Since supplier capability is at the center of any acquisition or outsourcing decision, it is important to find out in advance if the contractors that comprise the supply chain possess all of the capabilities required to do the work.

Overall capability is usually demonstrated by the supplier's past history with similar projects as well as its documented ability to adopt good software engineering practices. The presence of a fully auditable, commonly accepted model of best practice shared between customer and supplier helps to cement that assurance.

Obviously, a big part of ensuring trust relies on the ability of the supplier to guarantee that it can deliver on cost, timeline, and integrity commitments. The problem is that, given the complexity of most ICT projects, it is difficult for any supplier to provide that sort of guarantee. According to Watts Humphrey, the three variables that serve as the basis for trust in business are history, understanding, and awareness. All ICT organizations have difficulty assuring any of these three factors since companies will normally deal with each other electronically, over continents and oceans. Therefore, unless the acquirer and supplier have done business with each other before, those companies often have little basis for gauging performance. Additionally, even if two companies have experience with each other's work, there is no guarantee that a customer can rely on similar results in a succeeding project, given all of the factors involved in technology product manufacture.

ICT products are developed through a global supply chain, and the goal is to supply a product or service through coordinated work involving several organizations. The problem is that ICT supply chains produce products that are either abstract, like software, or so infinitesimally complex that they cannot be overseen and controlled by conventional means.

Not only do we have an increasingly global interdependent supply chain, the complexity of modern ICT products demands a myriad of capabilities that are increasingly dependent on globally sourced ICT products. That creates a situation where attackers from a nation-state, terrorist, criminal, or rogue developer might be able to gain control of systems through supply chain opportunities and intentionally implant logic or create unintentional vulnerabilities that can be maliciously exploited.

The acquisition of ICT is a strategic business process, not a discrete managerial activity. That process is built around a logical sequence of seven managerial phases. The relevant activities in these phases need to be practiced as a continuous process throughout the operation of the sourcing and procurement function. The process is also dynamic in that it adjusts its component actions and tasks to changes in the technology or risk picture, as those occur in the day-to-day operation of the business. Naturally, the primary cause of that adjustment will be the appearance of new technology or threats.

ICT acquisitions are planned and deployed by means of a strategic management function. And the ICT acquisition operation is always maintained through a set of comprehensive and systematic policies designed to ensure optimal visibility and control. The ICT acquisition process is meant to be carried out as an organizational control process, performed on a no-less rigorous basis than corporate financial control. That means that all relevant acquisition process documentation should be maintained in auditable condition.

ICT SCRM is founded on the creation of an explicit management control structure and a concomitant set of controls. A single coordinated framework of controls is necessary, because business has developed a growing dependence on COTS technology, and with that increasing reliance comes the increased need to ensure that the technology being purchased is properly secure. Thus, as business organizations grow and diversify, progressively more powerful means of ensuring trust up and down the supply chain are required.

In day-to-day application, this means that it must be possible for companies to assure every product they source; this is a particularly difficult task to perform, given the layers of complexity involved in a global supply chain. Most product creation work is done in multilayered, multivendor, and even multicultural team settings, and all of the sourcing for the products integrated into a final system must be fully coordinated and controlled. Coordination and control of complex work necessitates a communal, shared point of reference that will allow the managers in the customer organization to benchmark the activities up and down the supply chain against best practice. That common point of reference is embodied in a security governance framework.

There are five processes that underwrite the development of a supply chain assurance solution to any given acquisition. The first of these processes is *scoping*. This term describes the intentionally planned and systematically executed process of establishing the boundaries and criteria of any sourced acquisition. In some respects, this is the most critical step, since the depth and actual links in the supply chain will determine the form and extent of the rest of the solution. In the real world, that implies a conscious balancing act between the need for trust in a product and the resources that are committed to establishing it.

Obviously, anything can be assured if enough money is thrown at it. However, no organization has the money to effectively put a cop on every street corner when it comes to policing a retrospective supply chain. So, a deliberate process has to be undertaken that balances deployment of the controls to assure against supply chain risks and the likelihood and material consequences of the threat space. Factors that might enter into this process include considerations such as the level of criticality for each product and the degree of trust and concomitant assurance required for that product.

The second SCRM process is *threat identification,* the organization's understanding of the nature of the threat environment, both internal and external. Because of the invisibility and complexity of ICT products, assessment is a primary player in any ICT SCRM process. Assessment generally involves identifying gaps in product purchasing performance and then aligning current SCRM practices against some sort of best practice reference model, which is usually embodied by the expedient of a well-defined and commonly accepted standard. The substantive processes that are then deployed reflect strategic SCRM best-practice models, such as ISO 28000, ISO 31000, and NIST-IR 800-161.

The *risk assessment* then evaluates the impact of the identified threats on the proposed purchase and the organization's overall system assets. A particular threat may not necessarily have much negative impact on the system within a given organization's environment. So, the entire set of identified hazards is examined in order to differentiate those threats that would create specific and undesirable impacts.

These impacts are then carefully assessed with respect to their short- and long-term resource implications. Those assessment ratings are then recorded and prioritized so that the threats with the most critical impacts are dealt with first.

The next principle is *design*. The outcome of the design process is a complete and proven effective day-to-day set of practices that address each issue identified in the risk assessment and fulfill all known constraints in concrete terms. All designs exhibit common characteristics. They are complete in the sense that they encompass the entire solution. They are correct in that all elements of the solution that logically should be present are provably there. They are understandable in that they communicate the form of the solution and that all of the management control elements in the design are traceable to the issues raised in the risk assessment.

In addition to direction, the design must provide a clear reflection of the anticipated degree of integration. Technical elements must be integrated at their interfaces, to promote the most efficient interaction. Business elements have the same requirement at the process, activity, and task levels. All of the necessary relationship issues must be identified and resolved in the design. In addition, as the name implies, along with the design, there is always some sort of formal planning process involved as well.

Because a plan provides the explicit direction for the integration process, it is the essential end product here. In the case of a small business, this might be a relatively trivial documentation item, a memo of agreement, for instance. However, where the security solution is very large or complex, there is an implicit element of long-range planning, and the result is always a detailed plan.

The largest principle, in terms of the actual time spent, is *integration*. This concept might be just as appropriately termed "realization," or "implementation" because its product is the actual, substantive supply chain-based product management process. In the case of SCRM, the general goal is to ensure that the means employed to identify and provide the requisite processing and storage embodies the proper set of controls.

The process logically involves the gathering of all of the information necessary to identify and categorize each of the supplier resources that will fall within the system. Essentially that involves the identification, labeling, and valuation of each and every one of the organization's ICT assets and the formulation of these into a comprehensive baseline inventory of products and their suppliers. In many respects, this part of the process resembles the organizational asset identification activity that would drive any fiscal accounting or physical inventory process, in the sense that individual categories of ICT product are simply differentiated and accounted for in a permanent ledger along with the sources of each product. This baseline ledger is maintained under the dictates of rigorous management control.

Once all of the organization's information assets have been identified and arranged into a catalogue of coherent baselines, the *next step* is to determine their current protection requirements. The purpose of this activity is to define the exact assurance needs for each of the individual items in the product/supplier baseline.

The *next three steps* are really aspects of the same process, which is the actual establishment of the day-to-day control system. Once all of the elements of supply chain risk and concomitant resource commitments are understood, it is time to create and array the initial set of controls for each product/supply chain the organization chooses to manage. This amounts to an iterative control validation and enhancement process.

The *final step* in this process is measurement. In a technical setting, this might be called a metrics program. Because ICT and network products constitute a virtual asset, measurement is the fundamental element that enforces management control. Quantitative assessment of the security functioning of a given supply chain, in the form of such measurement items as breaches, breakdowns, or identified flaws and/ or malware, helps decision-makers evaluate the ongoing performance as well as assign accountability.

The recommendations that are embodied in a best practice standard are meant to provide a concrete basis for characterizing a wide range of abstract assurance functions, including intangibles such as product information flows. In concept, the recommendations of a best practice standard are not standalone elements. They actually substantiate numerous facets of the day-to-day assurance operation that taken together constitute aggregate best practice.

An organization must originate the policies to fulfill its aims and purposes within its particular culture. This is embodied through a governance approach that centers on a coherent framework of selected controls. This governance process requires that there be some kind of standard point of reference to coordinate all security activities that will be carried out up and down the supply chain. This is why every one of these approaches requires a standard model of expert best practice. The rest of this book will be devoted to that discussion.

Key Concepts

- The SCRM process has many dynamic facets.
- SCRM for ICT centers on devising tangible means to identify and control risks up and down the supply chain and to ensure fundamental trust in a product.
- Supply chains are often not known, or well defined, in the product development process because of the existence of outsourcing.
- It is necessary to inventory and label all risks to all product elements in order to control them.
- The procurement process should be based on information about product provenance and supplier trustworthiness.
- In order to be effective, the overall SCRM process has to be planned and coordinated as a strategic process.
- Proper SCRM involves deploying and then maintaining an appropriate baseline of technical and managerial controls to mitigate risk.

- Effective SCRM ensures trust among suppliers and customers.
- Because it represents a change in routine practice, supply chain risk governance schemes require formal organization and executive sponsorship.
- Supply chains can be intentionally compromised by an adversary in order to promote its goals.
- Standard models of best practice for supply chain risk are important roadmaps for organizations to follow.
- The NIST 800-161 Standard is the first standard model aimed specifically at specification of SCRM best practice.

Key Terms

availability: a system or organizational state in which all necessary information is accessible at the time it is needed

compromise: a breakdown in organizational control leading to breakdown of a supply chain

confidentiality: a system or organizational state in which all relevant information is protected from unauthorized access

controls: technical or managerial behaviors that are put in place to ensure a given and predictable outcome

counterfeiting: the insertion of a component, or the sale of a product, that has not been produced by the manufacturer of record. This can lead to unexpected failures of components and products

countermeasures: technical or managerial actions taken to prevent supply chain failure either at the individual unit level or for the product as a whole

governance framework: a comprehensive set of standard activities intended to explicitly define all required processes, activities, and tasks for a given field or application

information governance: is the process of identifying a set of multi-disciplinary processes, procedures, security controls, and the management of information to support the operation of the organization in a holistic approach.

integrity: a system or organizational state in which information can be shown to be accurate, correct, and trustworthy

malicious code: small program intentionally inserted in a product to perform unwanted or harmful actions

National Institute of Standards and Technology (NIST): the body responsible for developing and promulgating standards for federal programs and Federal Government agencies

organizational governance: a condition that ensures that all organizational functions are adequately coordinated and controlled by policy, typically enabled by strategic planning

sourcing: the organizational process of product acquisition either through a development process or by the purchase of a COTS solution

strategic planning: the act of translating an organization's intended direction into specific steps along a particular timeline; strategic planning affects the entire organization for a significant period

References

General Accountability Office, IT supply chain: National security-related agencies need to better address risks, GAO-12-361: Published: March 23, 2012.

General Accountability Office, 2015 Annual Report: Additional opportunities to reduce fragmentation, overlap, and duplication and achieve other financial benefits, GAO-15-404SP: Published: April 15, 2015.

Humphrey, W., *Managing the Software Process*, Addison-Wesley, Reading, MA, 1989.

International Standards Organization (ISO), ISO/IEC 12207: 2008-Systems and software engineering—Software life cycle processes, ISO, 2008.

International Standards Organization (ISO), ISO 27001, Information technology—Security techniques—Information security management systems—Requirements, ISO, 2013.

International Standards Organization (ISO), ISO 31000, Risk management—Principles and guideline, ISO, 2009.

National Institute of Standards and Technology (NIST), NISTIR 7622, Notational Supply Chain Risk Management Practices for Federal Information Systems, NIST, 2012.

National Institute of Standards and Technology (NIST), SP 800-37 Revision 1, Guide for applying the risk management framework to federal information systems, NIST, 2014.

National Institute of Standards and Technology (NIST), SP 800-161, Supply chain risk management practices for federal information systems and organizations, NIST, 2015.

Privacy Rights Clearinghouse, *Chronology of Data Breaches Security Breaches 2005—Present*, PRC, San Diego, CA, 2014.

Chapter 2

Building a Standard Acquisition Infrastructure

At the conclusion of this chapter, the reader will understand the following:

- The structure of the ISO/IEC 12207 standard and its relevance to supply chain risk management (SCRM)
- Why an Agreement process is necessary to maintain a set of formal controls for managing supply chain relationships
- The underlying activities and tasks defined in the ISO/IEC 12207 Acquisition process
- The underlying activities and tasks defined in the ISO/IEC 12207 Supply process

The challenges in which SCRM affects the underlying system/software life cycle processes of an organization necessitate factoring in a standardized approach to overcome those challenges and build SCRM into the common life cycle processes that currently exist within an organization. Recall that the key ingredient is that SCRM be built into a well-defined process and provide capacity for improvement. To that extent, we suggest that optimally SCRM be inclusive to an established standardized life cycle framework.

For many years, information and communication technology (ICT) organizations have implemented a standard that is relevant to the definition of a robust acquisition assurance infrastructure at the conceptual level. The customer–supplier process defined in the "Agreement" processes of the ISO/IEC 12207:2008 *Systems and software engineering—Software life cycle processes* provides appropriate activities and tasks that, if implemented properly, will steer an organization into the direction of achieving the underlying requisites of SCRM. ISO/IEC 12207 has been widely accepted for over 20 years as the authoritative definition of what, at a minimum, must be undertaken for proper technology acquisition. These standard recommendations can then be tailored into a specific process for any given organizational application.

Thus, the need for a single fully defined infrastructure is a precondition for the definition of the body of knowledge for secure SCRM. In this chapter, we will begin with a discussion of the generic structure of the ISO/IEC 12207 standard. Upon understanding the ISO/IEC 12207 standard at a high level of abstraction, we will be able to "zero in" on the standard's Agreement process, which contains the activities and tasks that directly impact the ability to build an effective acquisition infrastructure that integrates the necessities of SCRM. In later chapters of this book, you will learn about the techniques for implementing the recommendations of NIST IR 800-161, *Supply Chain Risk Management Practices for Federal Information Systems and Organizations* within the larger ISO/IEC 12207 Agreement process.

ISO/IEC 12207

It was not until 1995 that a single standard allowed ICT organizations to organize, coordinate, and establish best-practice activities throughout the ICT life cycle. Then ISO/IEC 12207 (1995) and its American counterpart IEEE 12207.0 (1998) were developed to satisfy that requirement. The premise of that standard was to describe a complete set of practices for software work spanning the entire life cycle from initial product planning to retirement. Included in those standard practices were the Agreement processes that aimed to specify the standard activities for acquiring a correct ICT product or service. To that extent, 12207:1995 was the first standard to effectively integrate the diverse and disparate collection of industry best practices for a customer–supplier relationship into a single recommended approach to effective product procurement.

By implementing the recommendations of 12207:1995, organizations were able to effectively combine a set of procurement best practices into a single uniform approach. Further, the best practices could be tailored to specific ICT circumstances as a means of controlling all aspects of the product acquisition life cycle. The standard was developed in such a way that the itemized processes constituted a complete set of activities and tasks applicable to all development, operation, and maintenance situations. ISO/IEC 12207 is the comprehensive framework that allows enterprise architects to decompose a standard life cycle process from a generic set of management activities down to their instantiation in the form of concrete everyday tasks. The tasks are then tailored by identifying unique project issues, problems, and criteria and documenting adjustments from the standard definitions. Subsequent tailoring was carried out by decomposing process components to their logical level of expression in work instructions and other required processes and then implementing those practices in day-to-day situations.

From 1995 to 2008, the ISO/IEC 12207:1995, *Information technology-Software life cycle processes* standard could realistically be considered one of the most successful international standards ever written. It provided recommendations that gained universal adoption throughout the industry as the model for a well-defined ICT life cycle. However, as technology continued to evolve, the scope of business practices had changed by the early 2000s. So, it was increasingly evident that the recommendations of the 1995 standard were lacking and the model needed to be updated.

The Joint Technical Committee, ISO/IEC JTC 1, is the ISO committee charged with the responsibility of revising the ISO standards pertaining to information technology. Further, 12207 is ICT related; therefore, the revisions to the 12207 standard were made by JTC 1's Information Technology Subcommittee SC 7 for ICT and Systems Engineering. The most significant charge of the SC 7 subcommittee was to integrate the concepts, terminology, and framework of the ISO/IEC 12207:1995 standard with the organizational and project processes of the ISO/IEC 15288:2002 standard for the system life cycle.

Since ISO/IEC 15288, *Systems and software engineering—System life cycle processes*, is referenced in several of the NIST Special Publications related to risk management, it is worthy of mention here. Both the ISO/IEC15288 and the ISO/IEC 12207 standards were being revised at the same time. Therefore, SC 7 effectively refined a model originally designed to address the software aspects of life cycle activity into a model that was fully aligned with concepts and processes for complete ICT product and service delivery at the organizational level. The resulting version of ISO/IEC 12207, published in 2008, encompasses a larger set of project-oriented processes aimed to provide a comprehensive characterization of the environmental elements within which ICT products are developed. In essence, ISO/IEC 12207 and ISO/IEC 15288 are so tightly integrated that they provide one in the same in terms of life cycle process requirements. It is for that reason we use the ISO/IEC 12207 standard as a basis for our discussions in this book.

The original ISO/IEC 12207:1995 was composed of five primary life cycle processes, eight supporting life cycle processes, and four organizational processes. The 12207:2008 version incorporates a much larger set of processes into two areas of common focus: system context processes and ICT-software specific processes. The system context process section of the standard contains most of the processes that originated in ISO/IEC 15288. These processes are grouped into four logical areas: agreement processes, project processes, technical processes, and organizational project-enabling process groups. Each process group contains 3–11 life cycle processes that are further divided into a set of activities, and each activity is subdivided into tasks (Figure 2.1).

The processes from the original ISO/IEC 12207 are scattered across both areas of the new standard. However, most of the original ISO/IEC 12207 processes, particularly the supporting processes, are in the *Software Specific Processes* section. The two

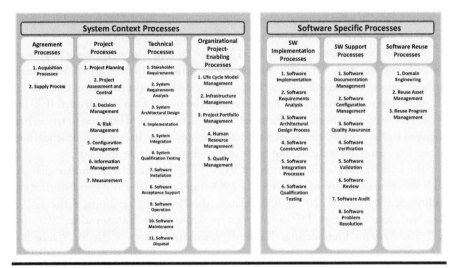

Figure 2.1   ISO/IEC 12207:2008 life cycle process groups.

ISO/IEC 12207:2008 processes that are an exception to the rule are the Agreement processes: Acquisition and supply (which are of interest to us in this chapter).

Since many products can be built from a single correct process, the creation of an enterprise-wide architecture from an ideal model of overall best practice is likely to solve many problems associated with improper management (throughout the supply chain) of the ICT process. In this respect, the ISO/IEC 12207:2008 standard provides the generic model that defines the ideal structure of the ICT process as a whole. It serves as a stable basis for defining a life cycle management framework that it is applicable to any form of ICT operation. Additionally, ISO/IEC 12207 provides a commonly recognized, worldwide basis for standardizing terminology and processes to effectively manage any software or ICT development, sustainment, or Acquisition process. The template of activities and tasks provided by the standard is intended to be applicable at all levels of ICT operation, from the organizational level down through projects and all the way to the application development and maintenance stage.

12207 activities encompass all of the tasks pertaining to the management of the system life cycle necessary to establish the full scope of development, maintenance, and use of ICT products and services. The specific process, activities, and tasks itemized in the standard focus specifically on what is required to describe the artifacts of management and development pertaining to all ICT projects.

The ISO/IEC 12207 model alone does not provide sufficiently detailed guidance to make an organization and its supply chain manageable. As we mentioned previously, the processes within the framework still have to be tailored to fit each given situation. As you will see in subsequent sections of this chapter, the activities associated with customer/supplier acquisition are no exception. In other words, an optimum approach can be engineered top-down for each ICT project using the 12207 framework because the model embraces all possible forms of activity. The implementation strategy is always top-down, meaning that an explicit process model can always be constructed for any given project at any level of definition from the reference framework. This hierarchical approach creates a practical, top-down management process from policies formulated at the enterprise level through practical management procedures at the operational level.

Throughout our discussion, reference has been made to the word "tasks." Tasks are the explicitly defined set of work instructions for a particular role in the process. These task definitions are created to describe an organization's prescribed steps for carrying out any given activity. The work instructions comprise the tailored process model for a specific project environment. In operation, work instructions are at the other end of the management spectrum from the generic best practice specified in the ISO/IEC 12207 standard model. Work instructions are project specific and generally cannot be applied to another project, even if it is similar in scope and intent. Instead, the work instructions represent an organization's current best approach to executing the tasks required for a given project.

In sum, ISO/IEC 12207 provides the basis for defining the systematic activities, roles, and tasks of the ICT life cycle. Because those elements are defined in concrete terms, they can be easily evaluated by all levels of management and throughout the supply chain. The 12207 standard is applicable to every ICT organization and is particularly useful in the case of complex systems where SCRM considerations resulting from multitiered vendor arrangements and interoperable product lines are paramount and when the requirement for integration places exceptional demands for coordination of the process. By defining the best practices for each life cycle element, interruptions of the practical outcomes are minimized at all levels of work. Definition also helps ensure early risk detection and true product-in-process visibility.

ISO/IEC 12207 can be used in compliance situations either to dictate the execution of an individual project or to shape the process architecture throughout the supply chain. Compliance to 12207 is defined as the performance of all processes, activities, and tasks identified as appropriate in the tailoring process. The performance of a process or an activity is complete when all its required tasks are performed in accordance with established criteria and applicable requirements specified in the contract. However, compliance may be interpreted differently for various situations, including but not limited to compliance efforts at the organizational level, the project level, and when multiple suppliers are involved.

Agreement Processes: Overview

In order to facilitate such an acquirer/supplier relationship where SCRM practices are a priority, each party must engage in a predefined set of life cycle activities. The activities serve as a basis for assuring that the agreements established by the parties are upheld to the extent that a secure hardware, software, or service product is delivered.

The Agreement processes are the first set of System Life Cycle Processes provided by the ISI/IEC 12207:2008 standard, and for good reason. For all intents and purposes, the successful building of acquirer–supplier relationships serves as a foundation and basis for successful completion of *any* ICT system development project. Unlike the predecessor 1995 version of the 12207 standard, which separated the responsibilities of the customer and supplier into two primary processes, this first cluster of processes of the 2008 update combines them into a single category called the Agreement processes (6.1). These processes are particularly critical to ICT and SCRM because it is almost impossible to ensure the trustworthiness of an important purchase without a well-defined and rigorous process to guide the effort. Within the private sector or the national economy, the business missions, strategies, and defense rest on dependable ICT. Ensuring that the ICT that is purchased is acceptably secure is an important step in business and our everyday lives.

Every ICT manager who has engaged in building and maintaining customer–supplier relationships will admit to the challenges that they face. As we alluded to in Chapter 1, that is because the customer (acquirer) is in the position of purchasing a complex and sometimes intangible set of functions without any influence over how they are built or the security measures in place in the suppliers' development process. Therefore, SCRM begins to be evident when organizations utilize a set of procedures (such as those provided in the Agreement processes of 12207:2008) specifically designed to increase the acquirers, understanding and influence within the process. Moreover, those procedures provide an assurance measure/security control so that security is built into the final product. In providing the necessary security assurance, those procedures must guarantee that security considerations are a central part of the routine vendor selection, monitoring, and acceptance process. Likewise, in order to ensure their reliability, they should be well defined and consistent within the enterprise's organizational architecture.

The activities prescribed within both processes in the 12207 Agreement category provide the necessary tasks an ICT organization (acquirer or supplier) should perform in order to adequately manage the procurement of a system, software, or service product that has been determined as within the bounds of an ICT infrastructure. We discuss these activities within the context of this book because the Agreement processes are significantly relevant to individuals interested in enforcing the practices of SCRM within an organization, such that a structured and rigorous set of management activities and tasks is used to carry out the effort. As you will see in a later section of this chapter, the 12207:2008 activities specified for Acquisition (6.1.1) prescribe the practices to be performed when an organization procures a software system or service, while the Supply process (6.1.2) defines the obligatory activities carried out by the supplier. In short, by using the 12207:2008 standard Agreement processes, organizations can be confident that they are engaging in a detailed definition of customer–supplier relationship logistics directly

ISO/IEC 12207 Agreement Processes and Activities	
6.1.1 Acquisition Process	
6.1.1.3.1	**Acquisition Preparation**
6.1.1.3.2	**Acquisition Advertisement**
6.1.1.3.3	**Supplier Selection**
6.1.1.3.4	**Contract Agreement**
6.1.1.3.5	**Agreement Monitoring**
6.1.1.3.6	**Acquirer Acceptance**
6.1.1.3.7	Closure
6.1.2 Supply Process	
6.1.2.3.1	**Opportunity Identification**
6.1.2.3.2	**Supplier Tendering**
6.1.2.3.3	**Contract Agreement**
6.1.2.3.4	**Contract Execution**
6.1.2.3.5	**Product/Service Delivery and Support**
6.1.2.3.6	**Closure**

Figure 2.2   ISO/IEC 12207 Agreement processes and activities.

related to ICT procurement. However, that definition does not take risk management into consideration. As noted in the chapter introduction, risk management implications will be mapped back to the 12207 Agreement process activities in later chapters of this book.

Figure 2.2 provides a complete list of all activities prescribed for both of the 12207:2008 Agreement processes. Throughout the remainder of this chapter, we will take a detailed look at each of those activities, identify the tasks within them, and consider their implications to the overall approaches pursued by organizations to achieve quality and security benefits of SCRM.

Acquisition Process

Put simply, the Acquisition process defines those activities related to the customer role in purchasing system hardware, software, and services. The prescribed activities for acquisition are intended to convey the specific tasks to be performed by any organization that has identified the need to acquire a hardware or software system or other form of ICT service. Note that during the early discussions of this process, we use the term "generic" to take the process out of the defined context of the 12207 standard. We will later explain each activity of this process, prescribed by the standard, in detail.

Acquisition always operates in combination with the Supply process (6.1.2), which prescribes the supplier's responsibilities in providing that hardware/software product or service. The standard activities of the Acquisition process describe an ideal way to interact with a supplier, to ensure appropriate practices of SCRM. As you will see in a later section of this chapter, the Supply process delineates the ideal way to deliver a product or service. These two processes together are the basis for formulating an ICT service delivery or system purchase contract. The 12207 standard qualifies its intent by pointing out that its Agreement processes were derived as a subset of the Agreement processes specified by the ISO 15288 standard, which is suggestive of the tight integration between the two standards that we mentioned earlier.

In the same vein as the acquisition activities prescribe ideal steps in interacting with the supplier, in a more general sense they convey the ideal steps an organization would take to acquire any ICT system or service product. Thus, to suggest that the activities only aim to define the customer–supplier relationship would not provide adequate context to the process outcomes defined in the standard. The underlying goal of the Acquisition process is that the customer purchase fundamentally correct ICT products and services. The process begins with the identification of customer needs and ends with acceptance of the product or service as defined in the contract. The Acquisition process itself is a strategic function in that it is established and maintained through formal planning. Planning for each acquisition project takes place within the framework of practices established in the overall strategic planning activity.

Without disregard for the established outcomes identified in section (6.1.1.2) of the 12207 standard, the main objective of the Acquisition process can be summed up in the statement that its underlying purpose is to ensure that the right vendor is chosen. Since vendors supply a product or service, both the acquiring and supplying organizations must know the specific requirements for what is being delivered in order to make an intelligent decision. Therefore, the first step of the more generic sense of the Acquisition process is to define and document all pertinent product requirements, in addition to the criteria the organization will use to make a judgment decision as to whether those requirements were met according to definition. The outcome of this generic requirements definition activity should be a legal contract that designates all required roles, responsibilities, and obligations of both parties in the acquirer/supplier relationship.

Likewise, the contract should provide parameters for the appropriate monitoring and control mechanisms that must be in place in order to ensure that the supplier upholds the terms and conditions of the contract. Once the contractual terms have been satisfied, the generic Acquisition process then ensures an efficient transfer of the product from the vendor to the customer.

Acquiring a commercial off the shelf product (COTS) or outsourced service is often a beneficial option over making it or providing the service yourself. The reason for that is due to the increased business advantages provided by COTS over a custom product or in-house operation. Further, purchased products are normally less expensive and are immediately available. However, COTS products do come with problems, many of which involve the anonymity of the supply chain. Therefore, an effective process of risk identification and risk management is an essential part of the Acquisition process, even as early as the business justification and requirement definitions phase for a COTS product. As you have learned in Chapter 1, SCRM can be easily summed up as a process used to ensure that an acquired product satisfies the basic requirements for integrity and safety. Nevertheless, consideration of risk is important throughout the process because specific actions to ensure product quality and security have to be designed and implemented from the start. A proper understanding of risk helps the purchaser make better decisions about the level of investment needed to address all of the known risks.

On the other hand, outsourcing considerations also represent an important issue in the contracting process. Therefore, it is crucial for the acquiring organization to do the analysis needed to understand every aspect of any outsourcing decision. The analysis must be completed before an organization decides whether to contract out the work or service. First, supplier capability is crucial to any outsourcing decision, so the acquiring organization must determine, in advance, if the contractors in the supply chain have the capabilities required to do the work or provide the required service. Moreover, it is the responsibility of the supplier to prove that it can develop and integrate a secure product or service. Most often, capability will be determined by researching the history of the supplier with regard to delivery of similar products or services as well as a documented ability to adopt good ICT engineering practices and employ trustworthy subcontractors.

Understanding the variability that exists between subcontractors, it is important for the acquiring organization to explicitly state which elements of the proposed work its subcontractors will *not* be allowed to perform. Being up front with such statements affords an opportunity in which the acquiring organization and the prime contractor can make specific determinations about how to apportion the work as well as manage and support the subcontractors. The assignment of work elements to subcontractors is then stipulated in the contract along with all outsourcing decisions.

Contractual consideration must also be given to the degree of foreign influence and control that might be exercised over the product or service. If a COTS product or outsourcing comes from an organization within a nation not necessarily friendly to the United States, the product or service might not be considered trustworthy. Therefore, the degree of foreign influence, control, or ownership of a contracting organization has to be determined before additional steps are taken to manage risk.

The final generic consideration that needs to be made about the Acquisition process, before we move into the detailed activities defined by 12207, deals with competitive pressure. Competitive pressure is a vital element in the Acquisition process because it is the most effective practice used to get the best possible deal. One way that an acquiring organization can maximize competitive pressure is through a bidder's conference, which puts potential suppliers in the position of competing for the organization's business. Ultimately, the competition drives down the eventual price and puts suppliers on notice that they must submit their best bid.

Acquisition Activity: Acquisition Preparation

As is characteristic of all 12207 processes, the Acquisition process begins with a formal initiation activity. Those familiar with ICT industry-adopted life cycle methodologies will be interested to note that the first several tasks defined in this activity parallel the first few steps of any of those methodologies. Figure 2.3 provides a summary of the tasks of this activity.

Before delving into the detail of this activity's tasks, it is important to make a few observations. In addition to the preparation tasks taking on the characteristics of many ICT system life cycle methodologies, the tasks may not necessarily be

ISO/IEC 12207 Acquisition Preparation Tasks	
6.1.1 3.1.1	"Concept of Need" Developed
6.1.1.3.1.2	System requirements defined and analyzed
6.1.1.3.1.3	Requirements specification prepared by the acquirer or assigned supplier
6.1.1.3.1.4	If a supplier is used to prepare requirement specification, an approval authority is sought
6.1.1.3.1.5	The standard's Technical processes are used to establish the system requirements
6.1.1.3.1.6	Consideration is given to acquisition of any system requirements
6.1.1.3.1.7	Off-the-shelf software chosen for acquisition is matched to standard conditions
6.1.1.3.1.8	Acquisition plan is prepared and executed
6.1.1.3.1.9	Acceptance strategy and conditions defined and documented
6.1.1.3.1.10	Acquisition Requirements (Request for Proposal) are prepared
6.1.1.3.1.11	Acquirer determines which of the other 12207 processes are appropriate and in need of tailoring. Determination is made and documented in the acquisition document as to which processes are performed by the supplier or further down the supply chain.
6.1.1.3.1.12	Contract milestones are determined and documented in the acquisition document
6.1.1.3.1.13	Acquisition requirements are given to the organization chosen to perform the acquisition activities

Figure 2.3   ISO/IEC 12207 Acquisition preparation tasks.

performed by the acquiring organization. The 12207 standard defines the tasks of this activity in such a way that some of them may be performed by a supplier (subcontractor) or even their subcontractors. This point becomes evident beginning with task 6.1.1.3.1.3 and is built into the task definition in 6.1.1.3.1.4. The formal definition of the standard task 6.1.1.3.1.11 states:

> The acquirer should determine which processes of this International Standard are appropriate for the acquisition and specify any acquirer requirements for tailoring those processes. The acquirer should specify if any of the processes are to be performed by parties other than the supplier, so that suppliers may, in their proposals, define their approach to supporting the work of other parties. The acquirer shall define the scope of those tasks that reference the contract.
>
> *(International Standards Organization, 2008)*

One logical interpretation of this task would suggest that not just one or two, but any of the Acquisition process tasks could be outsourced. Thus, it is easy to see that the implications of SCRM become a priority very early in the Acquisition process. Nevertheless, if any of the Acquisition process tasks are outsourced, it is generally the task defining and analyzing the system requirements. That being said, it would be realistic to conclude that a Request for Proposals (RFP) would need to be prepared and disseminated to potential suppliers as early in the life cycle as the approval of the concept of need and be based primarily on that document in combination with the Project Management Plan (PMP).

Concept of Need

Many textbooks and the ICT literature describe the concept of need as the statement justifying the predetermined feasibility of an ICT project based on the findings of a larger scope Business Process Reengineering (BPR) effort. Since this book addresses the risk management practices in customer–supplier relationships, it is useful to discuss this task within the context of developing a concept of need from the assumptive perspective that some, if not all, of the system hardware, software, and services are provided through the supply chain. In a large organization, a request to purchase can originate from any level of the company. So, it is difficult to ensure that every request is justified. Therefore, the first practical step in the ICT Acquisition process is to prepare a fully documented concept of need, which provides justification for a proposed system acquisition. In that respect, the practical business advantages and risks of buying that new product or service must be explained before moving forward into further tasks of the Acquisition process.

Documenting the need to undertake acquisition normally involves a specific statement of the business advantage of purchasing the product or service. Thus, the statement describes the business justification, which might be based on a business-planning project or on nothing more than a business decision by a corporate officer. Nevertheless, the 12207 Acquisition process requires that the originating party make the business case for the purchase before the processes can proceed.

Define, Analyze, and Document System Requirements

Once the justification has been made and an organization has considered all the factors of properly aligning the new product, the next step in the process is to define and analyze system requirements. Simultaneously, the customer (or contracted supplier) prepares a document that provides an explicit set of system-level functional and nonfunctional requirements. Those functional and nonfunctional requirements must characterize the prospective ICT product or service in detail. This standard document is called a System Requirements Specification (SRS). The SRS expresses each explicit product requirement in clear behavioral or functional terms. The process that underlies the development of the SRS analyzes all aspects of the acquisition, including the business, organizational, and user environment. In addition, the SRS considers all relevant safety, security, and other critical requirements in light of relevant design, testing, and compliance standards and regulations.

The requirements engineering process defines, analyzes, documents, and then ensures a suitable set of functions that will be embedded into the ICT product or service. To that extent, requirements engineering is a critical aspect of ICT acquisition activity because the requirements document provides the main point of reference for the supplier bidding process. Distinct qualitative criteria must be satisfied in order for a requirements document to be approved, including the clear statement of the qualification criteria for every functional behavior and nonfunctional quality that is specified in the requirements list.

Further, the main objective of requirements engineering is to provide a complete and correct description of the actions that the ICT service or system will perform. The description includes tangible qualities such as observable actions of the product or supportive tasks performed as a service, including inherent qualities such as security or reliability. In addition to categorizing all functional and qualitative requirements, the 12207 standard requires a description of all internal and external interfaces. This description has to be expressed in terms that are exact enough to ensure proper integration of all desired functions in the finished product. It is particularly important to assure seamless integration because risk management relies on the fact that the product or service is embedded without gaps that can be exploited by an adversary.

A full description of the target operational, physical, and environmental factors is necessary in order to understand the operating context of the system or ICT product. The operating context comprises the real-world conditions and demands that operate as constraints in either the construction or acquisition of the ICT product or service. Knowing these situational constraints helps eliminate unfortunate surprises that can drive up costs and slow production schedules.

The requirements document must sufficiently align all proposed ICT product and service requirements to their general system requirements. The system requirements provide detail on all relevant contextual factors, including business process and user needs. The system context is a vital component of the system requirements analysis and documentation task because it establishes the subsequent rigor of the system's confidentiality, integrity, and availability requirements. Thus, it is important to state each requirement in unambiguous terms.

The definition of system requirements is followed by the development of a specific set of lower-level functional requirements for each ICT product or service. For any project that follows a logical development path, the specification of lower-level ICT system requirements is always the next step. The reason for this is that individual ICT products and services operate within a system. Therefore, it is important to ensure optimum alignment between the contextual system needs and each ICT component. That fit is best ensured by a set of ICT requirements that can be mapped directly to the requirements of the larger context.

Developing a set of properly aligned ICT product requirements is not a task that should be approached haphazardly. In fact, as we mentioned at the beginning of this section, it often requires specialized expertise from a consulting organization. It is for that reason that the 12207 standard integrates supplier influence in many of the tasks within this activity. Nevertheless, regardless of whether the customer's organization or a contracted supplier does the basic analysis and requirements documentation, a senior manager from the acquiring organization must provide final endorsement that the requirements are correct.

Consideration for Acquiring System Requirements

Much of the decision-making associated with determining which (if any) products should be developed in-house or purchased off the shelf, and which product development or services outsourced, levels on the constraints imposed by "going it alone" or seeking supplier intervention and the amount of risk invoked in pursuit of either approach. It is not an overstatement to suggest that the considerations made in this task of the preparation activity have the greatest impact on the SCRM practices performed by the acquiring organization.

COTS and outsourcing considerations represent important issues in the contracting process, so it is especially important to do the analysis needed to understand every aspect of any outsourcing decision. The analysis must be completed before an organization decides whether to contract out the work.

First, supplier capability is crucial to any acquisition decision, so organizations must find out in advance if the manufacturers and suppliers all the way down the supply chain possess the capabilities required to provide the product or service.

Likewise, suppliers must be able to provide substantial proof that they can develop and integrate a secure product. Overall capability is usually demonstrated by the supplier's history with similar projects as well as a documented ability to adopt good ICT engineering practices and employ trustworthy subcontractors. If the acquiring organization cannot say with confidence that a trustworthy supply of a product or service is possible, then the decision must be made to do the work in-house or explore other alternatives for satisfying the associated system requirements.

Second, the variability of outsourced subcontractor capabilities must be considered. The acquiring organization should explicitly state which elements of the proposed work its subcontractors will *not* be allowed to perform. This allows the acquirer and the prime contractor the ability to make specific determinations about how to effectively and securely divide the work as part of management and support the subcontractors. The assignment of work elements to subcontractors is then stipulated in the outsourcing supplier contract along with all additional directives.

A third consideration that must be made is the degree of foreign influence and control that might be exercised over a particular product or service. If a product or service being considered for acquisition comes from an organization or nation that is not necessarily friendly to the United States, the product or service might not be considered trustworthy. Therefore, the acquiring organization must deliberate about the possible repercussions of contracting with a supplier with such foreign connections before additional steps are taken.

Preparation and Execution of the Acquisition Plan

The acquisition plan provides a formal approach to providing details that must be achieved within the Acquisition process for a given project. The plan also describes the relationship between a single project Acquisition process and the Acquisition process adopted by the acquirer at the organization level. At a minimum, the document includes all relevant criteria for making business decisions, such as financial and technical feasibility, contract milestones, audits and reviews, contractor performance, and budgetary and scheduling criteria. From an ICT product-engineering standpoint (Figure 2.4), the plan also specifies the following:

- Requirements of the system
- Planned employment of the system

Figure 2.4   Preparation of the acquisition plan.

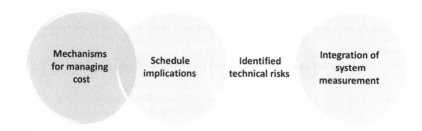

Figure 2.5   Criteria to include in acquisition plan.

- Type of contract(s) to be employed
- Responsibilities of each organization providing artifacts to the project
- Support mechanisms to be used during the project and throughout the operations and maintenance life cycle phases
- Risk implications and mechanisms for managing that risk

It is important to note that, in cases in which a supplier contracts with a subcontractor, it is the supplier's responsibility to coordinate the contract with the subcontractor. Additionally, the acquirer should establish and document a system measurement program. That program should be used to plan and track all activities and tasks across the life cycle (Figure 2.5). Criteria that should be included within the program include the following:

- Mechanisms for managing cost
- Schedule implications
- Identified technical risks
- Integration of system measurement with existing program management, programming, and other ICT functional areas pertinent to the project

Acceptance Strategy Definition and Documentation

Acceptance means that the supplied product or service has been tested to the extent that the supplied product or service contributes to the underlying definition of need set forth in the concept of need document, that the product or service adequately contributes to satisfying the requirements that meet that need, and that it can be supported by the acquiring organization. From a strategic ICT management standpoint, it is never too early to determine the criteria used for accepting an externally developed hardware or software product that will eventually become an artifact to the acquirer's ICT system or the service provided by a contracted supplier. The 12207 standard stipulates that during the preparation activity of the Acquisition process, the organization must begin developing a strategy for how acceptance will

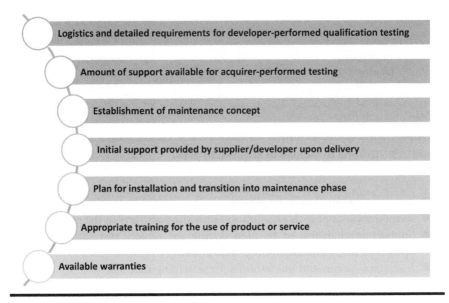

Figure 2.6 Criteria for acceptance strategy.

be determined (Figure 2.6). Such a strategy should include consideration of the following criteria:

- Logistics and detailed requirements for developer-performed qualification testing
- Amount of support available for acquirer-performed testing
- Establishment of a maintenance concept
- Initial support provided by the supplier/developer upon delivery
- Plan for successful installation and transition into the maintenance phase of the life cycle
- Appropriate training for use of the product or service
- Available warranties

Prepare Acquisition Requirements

The acquisition requirements (more widely termed RFP) is a document prepared by the acquirer that provides all of the criteria that adequately describe the product or service requirements and the communication of requisites put forth by the acquirer defining what will be required (of the supplier) to enter into a customer/supplier agreement.

It is worth emphasizing that an RFP is a formal business artifact. The aim of an RFP is to describe a product clearly and unambiguously to a prospective supplier.

The deliverable that underpins the description is a highly detailed SRS, which uses formal contractual terms to spell out the functional requirements needed to achieve a satisfactory solution.

Therefore, a good RFP must fully itemize the following:

- The functions that must be delivered in the finished product
- The criteria that the customer uses to evaluate and confirm whether the desired functions are present and correct
- Instructions for bidders
- Terms and conditions of a proposed contract between the acquirer and supplier
- Definition of control of subcontractors
- Technical constraints that may have been identified within the target environment

The main objective of RFP preparation is to afford anyone involved with the system an opportunity to contribute to its definition. Further, participation in the RFP development ensures both organizational and user buy-in when the system is delivered and installed. It is discouraging to mention, however, that the organization's technical staff often considers this step unnecessary. Nevertheless, the pragmatic world of business demands an open process involving all relevant stakeholders. That is because all ICT product and service specifications involve trade-offs and so a successful product must consider the needs of competing groups and satisfy each group's requirements.

Organizations do not need an RFP for everything they buy, but an RFP activity offers a number of benefits for any serious purchase. First, the process of preparing an RFP forces the organization's ICT, legal, and financial staff to analyze all aspects of the purchase. An RFP also puts a nontechnical manager in the driver's seat during negotiations with the supplier because it states the technical requirements in terms that a layman can understand. More importantly, a clearly stated RFP increases competitive pressure.

Acquisition Activity: Acquisition Advertisement

One very distinctive observation is that the engagement of the task within this activity begins a distinctive flow of alternating tasks between the standard Acquisition and Supply processes. Nevertheless, once the organization has defined its requirements for the new system or ICT service, the next step is to contact the people who can provide it. Those are the suppliers (ISO 12207, clause 6.1.2). The accepted way of initiating contact with a supplier is to issue the RFP created in the previous activity. Potential suppliers will respond with interest in the proposal through a prepared bid (activities: 6.1.2.3.1 and 6.1.2.3.2).

There is no single approach for advertising the RFP to suppliers. Earlier in the chapter, we mentioned a competitive pressure approach to attracting suppliers.

It should be noted, though, that it is more often the case that the acquirer approaches potential suppliers on its own, rather than the other way around. Regardless of the approach, the 12207 makes an important note that this task is likely to include supply chain management partnering. Such a practice involves exchanges of information with related suppliers and acquirers to reach a collaborative approach to common technical and commercial issues. Thus, SCRM must be in place to protect the shared information.

Acquisition Activity: Supplier Selection

Earlier in this book, we emphasized the importance of trustworthiness between the customer and supplier in a supply chain relationship. At the point that the acquiring organization performs the tasks of this activity, that trustworthiness must be unequivocally evident in materials used to make the selection. The tasks of this activity can be likened to an employer selecting a new employee to perform an organizational responsibility. The employer agrees to provide the resources needed to perform the job function. The organization makes the appropriate selection of an employee who can efficiently and effectively perform the role and can be trusted. Figure 2.7 provides a summary of the tasks of this activity.

As we explained in the overview for the Acquisition process, the acquiring organization is responsible for developing a logically derived, formal model for supplier selection. This model usually includes the methodology and criteria used to evaluate the proposal. Moreover, the weights and methods employed to assess compliance are usually included in the evaluation model.

The actual assessment and selection of the supplier should be based on its ability to deliver the system, ICT product, or service as indicated in the documentation provided in the RFP. We mentioned in our discussion of the 12207 standard that each defined activity is tailored to a given project. Thus, the acquiring organization may seek the assistance of other parties, including potential suppliers, before awarding the contract. By rule, all final decisions in tailoring are made by the acquirer.

The process for evaluating each supplier's proposal centers on the determination of whether all requirements described in the RFP are documented, traceable, testable, feasible, and consistent with the needs of the acquiring organization.

ISO/IEC 12207 Supplier Selection Tasks	
6.1.1.3.3 Supplier Selection Tasks	
6.1.1.3.3.1	Supplier selection procedure established
6.1.1.3.3.2	Supplier is selected based on the evaluation of the suppliers' proposals, capabilities, and based on the acquirer's acceptance strategy and conditions.

Figure 2.7   Supplier selection tasks.

The acquiring organization should also evaluate the supplier's capability to comply with the contract. The winning bidder is selected after appraising each supplier's ability to satisfy the RFP's terms and conditions. A prime consideration is the supplier's ability to comply with the standard's specifications for the Supply process (6.1.2). Once a winning bidder is selected, the acquiring organization and supplier sign the contract.

Acquisition Activity: Contract Agreement

The next major step of the Acquisition process is preparing the contract and engaging in negotiation with the selected supplier. It is critical that the contract address all known requirements of the acquisition project, not limited to just cost and schedule. It is also not uncommon to address pertinent legal topics such as usage, ownership, warranty, and licensing rights associated with the product or service being delivered. Once signed by both parties, the legal form of the contract is monitored and controlled through a formal change control mechanism. Assuming a life cycle process based on 12207, that mechanism is stipulated through the standards Configuration Management process.

The rationale for utilizing a change control mechanism is due, largely, to the fact that the contract is a living instrument. Often, changes crop up during project execution. Therefore, the contact must have a formal process for ensuring its continuing integrity in light of such changes. The roles and responsibilities for change management, as they relate to supplier contracts, must be defined as part of the configuration management plan. The stipulations outlined in the plan must address the following issues:

- Formality of the process with regard to contract change requests
- Methodology of communication to impacted stakeholders when changes occur
- Process for renegotiating the contract as a result of changes

The most common area for contract changes is the business areas, so an organization must put an effective policy in place to ensure the contractual integrity of any changes to project plans, schedules, budgets, costs, and security.

Common practice is for the acquiring organization to control the modifications that take place to the contract, that result from ongoing refinement of the product Acquisition process. Change is common and a process must be in place for negotiating the contract. Those organizations that have adopted the 12207 standard can implement the activities of the joint review process and problem resolution process. Under these two processes, contract changes are analyzed to determine their potential impact on project plans, costs, benefits, quality, and schedule.

The contract must address all requirements documented in the RFP, including all cost, schedule, and delivery concerns. Because the quality and security of

ISO/IEC 12207 Contract Agreement Tasks	
6.1.1.3.4 Contract Agreement Tasks	
6.1.1.3.4.1	The acquirer considers process tailoring requirements consistent with the selected suppliers' lifecycle process.
6.1.1.3.4.2	The acquirer prepares and negotiates a contract with the supplier that takes into consideration the identified acquisition requirements. The contract should also include criteria for proprietary, usage, ownership, warranty and licensing rights associated with off-the-shelf products.
6.1.1.3.4.3	The acquirer maintains complete control of any changes to the contract that may occur through negotiation with the supplier as part of a larger scope configuration control process.

Figure 2.8   Contract agreement tasks.

reusable and off-the-shelf components are essentially beyond the acquiring organization's direct control, the contract needs to address how to assure ownership, warranty, and proprietary, usage, and licensing rights associated with reused ICT products. Figure 2.8 provides a summary of the tasks of this activity.

Acquisition Activity: Agreement Monitoring

In this chapter, we are focusing on the activities and tasks of the 12207 Agreement processes. However, as you saw in the last section, each of the standard processes does not exist in isolation. There is a distinct dependency of one process on activities that exist in another. Generally, that dependency is on those processes that stipulate activities supportive in nature. This point is specifically true of the tasks that make up the acquirers' responsibilities related to agreement monitoring.

While discussion of processes other than Agreement is beyond the scope of this book, the 12207 standard stipulates that the acquirer should monitor the supplier's activities by employing the activities and tasks defined in the Software Review process and the Software Audit process. The standard continues by stating that the acquirer should supplement the monitoring with Software Verification process activities and the Software Validation process activities as needed. Finally, the acquirer must exhibit cooperation with the supplier, by providing pertinent information in a timely manner. Figure 2.9 provides a summary of the tasks included in this activity.

Describing the tasks of this activity in general terms, when the supplier is ready to deliver the completed system, ICT, or service, the acquiring organization must be prepared to conduct a formal acceptance activity that is usually itemized in great detail in the contract. Specific terms of the contract such as preparation of

ISO/IEC 12207 Agreement Monitoring Tasks	
6.1.1.3.5 *Agreement Monitoring Tasks*	
6.1.1.3.5.1	The acquirer must monitor the supplier's activities in accordance with the Software Review Process and the Software Audit Process. The acquirer must supplement the monitoring with the Software Verification Process and the Software Validation Process as needed.
6.1.1.3.5.2	The acquirer must cooperate with the supplier to provide all necessary information in a timely manner and resolve all issues.

Figure 2.9 Agreement monitoring tasks.

test cases, test data, test procedures, test environment, and the extent of supplier involvement should be defined to validate the product.

The acquiring organization must conduct acceptance reviews and testing of the deliverable using an approved life cycle process. The product is considered to be "accepted" when all reviews and tests are satisfied. Because the supplier's goal is to satisfy those conditions, the documentation for acceptance procedures is controlled by the supplier. To the contrary, the contract itself is controlled by the acquiring organization. After accepting the product, the acquiring organization must coordinate delivery of the product or service.

In the case of acquiring products, once they have been delivered, the acquiring organization must follow the prescribed acceptance procedures, auditing, and tests that were formalized in the contract. Once all tests and reviews have been passed, the acquiring organization accepts the product from the supplier and takes responsibility for the delivery.

Acquisition Activity: Closure

As is the case of the prepare activity that includes process initialization activities, another activity that all 12207 processes have in common is some form of closure. In the case of Acquisition, once the supplied ICT product or service has satisfied the conditions of the agreement and all open items have been settled, the acquiring organization concludes the agreement by rendering payment or other considerations and notifies the supplier that the contract has been closed.

Supply Process

Whenever an organization steps into the role of acquirer, it is doing so in pursuit of a relationship with a supplier or group of suppliers. In order to really have a grasp of the Agreement process and its impact on the overall scope of SCRM, it is necessary to understand the other side (the supplier's role). The two Agreement processes have a strong interdependence upon each other from the perspective of providing adequate and proper security functionality and risk management practice. Therefore, we cannot approach this discussion as simply a matter of viewing the Supply process as an adjunct to Acquisition. As you will see though the discussion that follows, the Supply activities are so tightly integrated with the activities of the acquirer that the two could feasibly be a single standard process. Such a unified purpose is important to the underlying principles of SCRM, and the broader acquirer–supplier relationship has been characterized under the umbrella of the broader named "Agreement" processes. The 12207 standard recognizes the dependencies that exist between the supplier and acquirer by defining the activities and tasks carried out by the supplier in reference to the Acquisition process discussed in the last section of this chapter. Moreover, the activities defined in the Supply process should be considered part of overall agreement on the requirements of the product or service. The Supply activities stipulated by the standard provide the best interpretation of an ICT industry set of best practices deemed necessary to provide an acceptable solution to an acquiring organization's needs.

To many, the Supply process is likely to appear very much like a project management activity. It coordinates and monitors the general function of providing a specified solution to a customer. It does not, strictly speaking, embody development tasks. Instead, it is chiefly composed of various planning and control functions. This very generic view of the process legitimizes its implications to the underlying practices of SCRM.

Contrary to many beliefs, the Supply process does not strictly involve development. To achieve its defined objectives, this process employs most aspects of the Development, Maintenance, and Operation activities. However, it is important that we emphasize that these are separate processes specified elsewhere in the 12207 standard.

The key point that organizations must keep in mind is that development, operations, and maintenance are separate life cycle processes meant to work complementarily with each other. The significance to our discussion, here, is that the activities in both Acquisition and Supply are not technical per se. Instead, they are intended to coordinate the more technical processes specified in the Development, Operations, and Maintenance activities of the 12207 standard.

Likewise, in the case where subcontractors are hired, the activities specified in the Acquisition process will also come into effect as a subset of Supply. In creating the overall process framework, the supplier should adopt two general organizational processes: management and infrastructure process development. Finally, as with Acquisition, the supplier must tailor individual project processes to the specifications of the 12207 Supply process using the Tailoring process defined in the annex of the standard.

Supply Activity: Opportunity Identification

Although not formally named such, the task associated with opportunity identification represents the initiation phase of the Supply process. As discussed in an earlier section, the task of opportunity identification is complementary to the acquisition advertisement activity of the Acquisition process. The "opportunity" can be recognized through a wide variety of mechanisms: electronic solicitation, word of mouth, cold calling by the supplier, or the more common means of awareness through previous business interactions with the acquirer. It would be through the later approach that the obstacle of establishing trustworthiness would become less of a concern since the acquirer and supplier have an existing working relationship.

This activity should not be approached haphazardly. In order to ensure a seamless coordination of supply and acquisition, the supplier must be selective in choosing the customers from which it conducts business. One vital element that should be considered is the extent to which the supplier processes blend with the processes of the acquirer. This becomes even more important when what is being supplied is critical to trust (e.g., cloud-based network infrastructure, storage capabilities, data backup, or security protection). Without trust and assurance of seamless supplier product or service delivery, the relationship between the parties is likely doomed from the start.

Supply Activity: Supplier Tendering

The Supply process continues when a supplier organization receives an RFP from a potential customer. Upon receipt of the document, the potential supplier must conduct a thorough review of the acquirer's RFP. This is done in order to identify all potential constraints within the problem domain, as well as to inventory potential solutions. The tasks of the supplier tendering activity are summarized in Figure 2.10.

The underlying objective of the RFP review is to define the product (or service) space. That space is made up of all of the possible solutions that will satisfy all

ISO/IEC 12207 Supplier Tendering Tasks	
6.1.2.3.2 *Supplier Tendering Tasks*	
6.1.2.3.2.1	The supplier conducts a review of the Request for Proposal
6.1.2.3.2.2	The supplier makes the decision to bid or accept the Request for Proposal
6.1.2.3.2.3	The supplier prepares a proposal in response the Request for Proposal

Figure 2.10   Supplier tendering tasks.

real-world constraints. Such constraints are likely to include all specified security considerations such as sensitivity and reliability. Clearly, trade-offs are a necessity in satisfying those constraints in order to provide the product or service since in most cases highly constrained sensitivity requirements will (for instance) impact the general availability and performance of the system.

Within the product or service space domain, any solution would be acceptable. Each would exhibit different external behavior, but all would satisfy all constraints. In simple terms, the outcome of the problem definition activity is a characterization of the range of practical solutions that would meet all of the known constraints (Figure 2.11). Sources of constraints may include the following:

- *Users* who have to decide what they need rather than what they want
- *Customers* who often stress financial factors over function
- *Technology* where the solution requires leading the target
- *Security* where there are specific security requirements (e.g., federal systems)
- *Laws* violations of which will produce unacceptable solutions (e.g., HIPPA)
- *Standards* if violated could produce unacceptable solutions

One aspect of this activity worthy of mention is that a normal result of constraint identification is the creation of a negative problem space, which happens when two requirements cannot fit the same project (for example, when the desired technology costs too much). So the real task of constraint identification is to evaluate possible trade-offs. Or in simple terms one constraint is relieved to accommodate another (for example, if we do not perform backups, then we can afford intrusion detection). Both the acquiring and supplying organizations' policy and procedure documentation must be consulted at this point to provide the acceptable

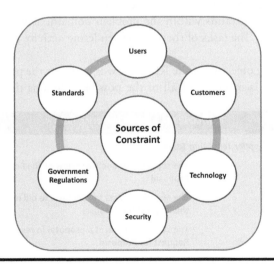

Figure 2.11 Sources of constraint.

terms for this analysis. In our example, it is commonsense that the trade-offs be heavily scrutinized. Once everybody is satisfied that the RFP is both feasible and cost justifiable, the supplying organization makes a decision as to whether to bid the contract.

If the supplier is confident that it can provide the service or create a product that meets customer requirements and satisfies all constraints, its management prepares a response to the bid request and then communicates a formal bid proposal to the acquiring organization. For all practical purposes, the proposal represents the formal response to the RFP and an indication that the supplier believes that it has the capability to deliver the product or service necessary to satisfy the acquirer needs. Additionally, the supplier develops a plan that specifies how it intends to tailor the recommendations of the 12207 standard to the identified requirements of the project. The plan is necessary because it acknowledges that the supplier's process is capable of complying with ISO 12207. Moreover, it could be considered the most important aspect of the practical work that has to be done to apply the process definitions of the 12207 standard. Even before a bid is submitted, the supplier must be able to demonstrate that it can execute the required processes, activities, and tasks of the standard within a defined process architecture. The response to the bid also specifies activities for the development, sustainment, delivery, and installation of the product or service. These activities all have implications for needed resources. Therefore, along with documenting how the supplier will carry out each recommendation of the standard, resource requirements must be determined.

As is the case in any ICT project, to make decisions about resources, a supplier must develop a project plan that ensures satisfactory delivery of the product or service within the constraints of the RFP. To help guide this plan, the Supply process typically employs many recommendations of the ICT product implementation processes. If subcontractors are hired, the recommendations of the Acquisition process are also used.

The point that needs to be made abundantly clear is that if we are looking for a point in the acquirer/supplier relationship where trust begins to develop to the extent that SCRM practices become effective, it is through the documentation provided by the supplier in this activity. Obviously, if an existing relationship already exists we can assume that trust has already been established. It is vital that both parties are aware of the processes in place and that those processes are defined to the extent that a secure product or service is delivered. Such a defined process is evident through the definitions provided by the 12207 standard.

Supply Activity: Contract Agreement

Upon acceptance of the supplier's proposal by the acquiring organization, a contract is negotiated and signed. A contract is a legally binding agreement; at a minimum, it should satisfy two specific rules. First, the contract will describe the entire

ISO/IEC 12207 Contract Agreement Tasks	
6.1.2.3.3 Contract Agreement Tasks	
6.1.2.3.3.1	The supplier negotiates and enters into a contract with the acquirer to deliver the proposed product or service.
6.1.2.3.3.2	Through a change control mechanism, the supplier may ask for contract changes.

Figure 2.12   Contract agreement tasks.

set of product or service functions in detail. Second, the contract will give both parties a clear means of determining whether a specified function or service has been provided (Figure 2.12).

The contract is the binding document between the acquirer and supplier participating in the procurement process, and it must legally define all requirements of the acquisition for all parties, including cost and schedule. Likewise, the contract must address such legal issues as usage, ownership, warranty, and licensing rights associated with the product or service being delivered. As a requisite to the contract's being the main mechanism expressing the product's or service's required tangible functions, those functions must be documented within each contract as externally observable behaviors. Moreover, the contract must specify a verification and validation process and a set of objective criteria that can be used to judge whether the behaviors have been correctly implemented in the product. Because technology and ICT products in particular are complex, the presence and execution of desired functionality are usually evaluated by proxy. Consequently, a contract will typically require objective, behavioral criteria that describe the functions they represent in unambiguous terms. These criteria allow the proxy to evaluate a complex product based on observable outcomes.

Over the life of the contract, it is not uncommon for changes to its language to be altered. The 12207 standard stipulates that the responsibility for ongoing maintenance of changes to the contract rests with the acquiring organization. If any alterations become necessary for successful delivery of the product or service, the new agreement is kept by the acquirer in the contract change control system. Likewise, maintenance of the contract under strict change control is a paramount priority from a security perspective. The reason for that high priority is due largely to the potential for undocumented requirements. Given the slim likelihood of being able to conduct a rational process if some of its elements are unknown, it follows that the logic of the need for contract control should be straightforward. Anyone who has ever worked within any facet of the development process would agree that strict control of changes is a difficult achievement (particularly the further along the project is, within the development process). That is why it is important to have a process in place that is understandable to all participants, exhibits risk management practices, and demonstrates management influence to the extent that the process is stringently followed.

Supply Activity: Contract Execution

At some point, the quibbling over legalities must be concluded and the hard work of providing the software or service that satisfies the acquirer's requirements must begin. It is within the contract execution activity that need becomes reality through a well-defined and managed supplier life cycle process. For those familiar with managing ICT processes, the tasks of this activity will seem familiar. It begins with a review of requirements, followed by the development of a PMP, continues by the selection of a managed life cycle process, and concludes when the process has been followed (all the way through qualitative testing) resulting in the desired product or service. Figure 2.13 summarizes the tasks of the contract execution activity.

Once the supplier has thoroughly reviewed and understands the acquisition requirements and has adopted a life cycle methodology that effectively supports the attainment of those requirements, the tasks of the contract execution activity proceed in an order very characteristic of any ICT project, beginning with the development of a project plan.

The project plan includes the documentation providing justified resource estimates and definitions of the scope and the extent of the acquiring organization's involvement at each stage of the life cycle. It is also the supplier's responsibility to develop a framework for quality, security assurance, and risk management implications from the supplier perspective.

It is important to note that the contents of the plan evolve directly from the 12207 standard processes. The 12207 standard stipulates that at a minimum, the following criteria must be addressed in the plan:

- Project organizational structure, including authority and responsibility of each organizational unit, interfaces with external organizations
- The monitoring and measurement environment (e.g., test library, test equipment, test environment, procedures, tools) for the project's development, operation, or maintenance activities
- A Work Breakdown Structure for each element in the deliverable, including the life cycle process and its associated activities, all specified ICT products and ICT services, and any items that cannot be delivered
- The selected approach to quality assurance for the ICT products or services
- The specific approach to managing safety, security, and other critical requirements for the ICT products or services; separate plans may be developed for safety and security
- Inclusion of Subcontractor management plans, defining an acceptable process for subcontractor selection and involvement

ISO/IEC 12207 Contract Execution Tasks	
6.1.2.3.4 Contract Execution Tasks	
6.1.2.3.4.1	The supplier conducts a review of the acquisition requirements to define the framework for managing and assuring the project and quality of the product or service.
6.1.2.3.4.2	If not stipulated in the contract, the supplier selects a life cycle model appropriate for the project. The life cycle must be mapped to the processes, activities, and tasks of the 12207 standard.
6.1.2.3.4.3	The supplier must establish requirements for the plans for managing and assuring the project and quality of the product or service.
6.1.2.3.4.4	The supplier must consider the options for developing the product or providing the service against an analysis of risks associated with each option.
6.1.2.3.4.5	The supplier must develop and document a project management plan based on the planning requirements and options selected in the previous task.
6.1.2.3.4.6	The supplier must implement and execute the project management plan developed in the previous task.
6.1.2.3.4.7	The supplier must develop, operate, and maintain the product or service in accordance with the processes of the 12207 standard.
6.1.2.3.4.8	The supplier must monitor and control the progress and quality of the products or services.
6.1.2.3.4.9	The supplier must manage and control the subcontractors on the project according to Acquisition Process. All necessary contractual requirements are passed down to the subcontractor in order to ensure that the product or service delivered to the acquirer is developed or per forms according to the prime contract requirements.
6.1.2.3.4.10	The supplier must establish an interface with the verification, validation, or test agent according to the contract and project plans.
6.1.2.3.4.11	The supplier must establish appropriate interfaces with all parties according to the contract and project plans.
6.1.2.3.4.12	The supplier must coordinate contract review activities, interfaces, and communication with the acquirer.
6.1.2.3.4.13	The supplier must conduct or support the informal meetings, acceptance review, acceptance testing, joint reviews, and audits with the acquirer as specified in the contract and project plans.
6.1.2.3.4.14	The supplier must perform verification and validation to demonstrate that the products or services and processes fully satisfy defined requirements.
6.1.2.3.4.15	The supplier must provide the acquirer, reports of evaluation, reviews, audits, testing, and problem resolutions as according to the contract.
6.1.2.3.4.16	The supplier must provide the acquirer access to the supplier's and subcontractors' facilities for review of products or services according to the contract and project plans.
6.1.2.3.4.17	The supplier must perform quality assurance activities.

Figure 2.13   Contract execution tasks.

- The plan for implementing, coordinating, and assuring the activities of ICT supporting processes, including the methods for assuring adequate verification and validation
- The plan for ensuring acquirer involvement in the process of developing or providing the solution. Such involvement includes the activities and tasks aimed at accomplishing joint reviews, external audits, informal meetings, reporting, modification, and change requests
- The plan that ensures user involvement by such means as formal requirements elicitation, prototype demonstrations, and product evaluations
- The risk management plan that must address mandated practices within the supplier organization and throughout the entire supply chain
- A formal specification of the security policy, generally provided by way of security plan and authorization package
- The mechanism employed to ensure adequate scheduling, tracking, and reporting project performance
- The plan to ensure the requisite level of education, training, and awareness for all project personnel

When all of the planning requisites have been defined, the supplier must then follow the most efficient approach to delivering the contracted product or service. The approach chosen must be selected based on factoring relative advantages and risks of each feasible option for delivering the product or service. Options for delivery include the following:

- Develop the ICT product or service using internal resources.
- Develop the ICT product or service through subcontracting.
- Utilize COTS ICT products from internal or external sources to satisfy part or all requirements.
- Use any combination of the three options, above.

Upon completion of the project plan and definition of the structure of work processes, the supplier moves forward by performing the work agreed upon in the contract. The first logical step in the process is to prepare a detailed plan of how the work will be done. Based on the stipulations of the 12207 standard, the project's technical work is performed by engaging the activities and tasks of the Organizational Project Enabling processes or activities of the system and software implementation processes. Depending upon the conditions of the contract, the supplier project work can include one or more of the following options:

- Develop the ICT product in accordance with the 12207 technical processes.
- Operate the ICT product in accordance with the 12207 operation process.
- Maintain the ICT product in accordance with the 12207 maintenance process.

The organizational management authority and project team member responsibility within the process are established once the overall approach is decided. Moreover, the decision regarding level of authority and responsibility must include all external organizations such as subcontractors and must take into consideration the directives defined in the organization's risk management plan. The task of defining assurance requirements is the critical first step in the planning process because product development takes place within those requirements. Likewise, assurance requirements must exist before any meaningful decisions can be made about practical aspects of production, including the environment, procedures, tools, equipment, and tests.

The supplier's main objective for the remainder of the project is to oversee and control the work as it progresses. That oversight responsibility includes the ensuring the performance of assurance work as defined in the quality assurance plan. The plan's oversight, control, and qualitative components are enforced throughout the term of the contract and throughout the life cycle. The 12207 standard emphasizes that oversight is a continuous operation that requires the supplier to exhibit complete control over the progress of its technical work as a means to control costs, schedules, and project status and to identify, investigate, and correct process-related problems as they occur. If a problem is identified, it must be submitted for resolution and formally closed before the ICT product can be considered acceptable.

Since it is the supplier's responsibility to create the quality assurance plan, it must take into consideration the extent to which the acquiring organization will be involved in the oversight process. Such involvement is normally evident through scheduled joint reviews, but audits can also increase the acquiring organization's awareness of the process. The caveat is that audits are quite expensive. Therefore, they are normally included as part of the language of the contract. Another important factor levels on the decision about whether or not informal meetings are appropriate between the acquiring organization and the various supplier types, such as primary contractors and subcontractors.

It should come as no surprise as you understand more about the supplier responsibilities for oversight and control that this management function plays a significant role in enforcement of risk management practices throughout the supply chain. Generally, there are several steps required to create a formal oversight and control management function that factors in risk awareness. The first step is initiation. Here, the objective is for the organization to define the leadership for security review, operational roles, and a formal organizational plan. Second, the organization must identify relevant review issues. This is often done in collaboration with organization line project managers from the supplier and acquiring organizations. The underlying objective of this process is for review managers and staff to identify and prioritize the key review issues.

Once review issues have been prioritized, the supplying organization can begin creating a generic review plan that includes the definitions for all pertinent audit and control activities.

Required standards and practices are also identified at this point, after which the review plan is integrated with the project plan. Once the review plan has been developed and integrated into the project plan, the organization can begin deploying procedures to guide the review process. This process normally involves training reviewers to perform reviews according to plan. Once there is confidence that the review personnel are prepared, the organization can begin implementing the review process. The implementation step involves assigning roles and responsibilities, developing a schedule, defining and performing monitoring activities, and reporting and resolving problems. Strategies must be in place for the review program to be evaluated periodically in order to determine whether it is performing effectively and as intended. A key point that must be made is that the reviewers must be able to provide sufficient, documented proof that the supplier's processes conform to the requirements of the contract. Moreover, the review must be able to confirm that the outcomes of those processes comply with established requirements and adhere to established and agreed-upon plans. If problems are identified during the review process, they must be documented and resolved before the review moves forward.

Due to the critical nature of the information of the review processes, in which documents are created providing a history of the product development process, that information must be recorded and stored. Likewise, since the results of reviews are important to the business function, records of their outcomes must be easily accessible to all supply team managers involved in acquisition of the product or service. There is a vital part of the review process that we have not discussed thus far—risk assessment reviews. Risk assessments provide the means by which organizations identify and evaluate all potential process, technical, and resourcing risks. The master review plan must provide a specific strategy for identifying and managing those risks and for addressing new or emerging risks. In turn, that strategy should map the organization's policies regarding risk acceptance and any formal procedures for reporting and addressing emerging risks. This brings us back to the main thesis of this book, which states that to ensure uniform application of the risk management process, the generic approach to risk must be standardized across all procurement projects and the entire supply chain.

Product or service integrity is ensured through the suppliers responsible for specifying a process to assure that integrity and security outcomes are stated in the contract. Such assurance is based on the overall risk management approach documented by that contract. The supplier must in turn implement a comprehensive assurance process that evaluates all identified risks for likelihood and impact. Once this assessment is complete and the supplier decides what risks will be mitigated and how that mitigation will take place, the supplier creates the plan for assuring the product or service and the defined project process elements.

Much of the 12207 standard stipulations for how reviews are conducted is provided within the standards Review Process. The activities within that process focuses on two kinds of assurance: product assurance and process assurance. Because reviews are primarily an oversight function, the initial goal of product assurance is to confirm that all work being performed complies with contractual requirements. However, the product must be reviewed regularly to ensure that it also satisfies all requirements specified in the contract.

Process assurance addresses defined requirements from a different perspective. The acquiring organization has a contractual right to a certain amount of transparency between customer and supplier. Customer visibility into the development process is ensured by the assurance process. Due to the procedural nature of the review, process assurance is the element that is potentially most interesting to the people responsible for the product's security. The product development process must comply with all project contracts, standards, and plans. Moreover, performance assessment metrics and evaluation criteria must be specified in the contract and then used for product and process evaluation. The assurance function must be able to adequately confirm that both the product and the process are compliant with all relevant standards. Assuring compliance with the contract requires rigorous monitoring of the project's general practices for product development, including the review and test environment and all practices related to process assessment. Likewise, since subcontractors are often involved in production, the process must ensure that all applicable contract requirements are passed down to the subcontractors and that their ICT products satisfy the requirements.

The key point to this discussion is that success of any project depends on the capabilities of the people who do the work. Reviews cannot make workers more capable, but they do enable organizations to find out whether workers possess the knowledge and skills necessary to meet project requirements. If workers need additional skills to meet requirements, reviewers have the responsibility of notifying the right people to ensure sufficient training.

Supplier oversight of project progress does not exist simply within organizational boundaries; the supplier is also accountable for strict oversight and control of any subcontracting organizations that work on any given project. The acquiring organization has the responsibility of ensuring that stipulations for managing subcontractors are provided in the contract. The supplier, in turn, is responsible for ensuring that subcontractors understand and execute all applicable contract requirements and conditions, including those for hardware, systems, ICT, and services. The supplier must ensure that processes are in place for conducting all contractually required verifications, validations, or tests of subcontractor work. Simply depending upon testing that may have been performed by the subcontractor is not adequate for compliance of the directives provided in the contract.

Management of the subcontractors also falls within the realm of responsibility of the supplier. It is imperative that the method used to select subcontractors be specified in the contract. Moreover, all involvement among subcontractors, the supplier, and the acquiring organization must be defined. The detailed acquirer/supplier/subcontractor relationship contract language must include all security assurance, verification, and validation activities, as well as the approach used for any third-party verification or validation agent. The 12207 standard stipulates that during project execution, the supplier is also responsible for communicating and enforcing all project requirements among subcontractors. The supplier must have knowledge of the Acquisition process to the extent that it is able to manage and control the activities of subcontractors in accordance with the process.

We have mentioned the assurance of subcontractor work through verification and validation. However, the supplier has the responsibility to implement any verification, validation, or test activities to development of services that they provide directly.to the acquirer and are required by the contract. In some instances, the supplier also finds it beneficial to work with independent verification, validation, or test agents, as specified in the contract and project plans. Nevertheless, regularly scheduled tests and reviews are important in supply operations because of the project oversight and control they provide. Testing and reviews are ongoing iterative activities that are meant to ensure the following:

■ The progress of technical work, contract performance, costs, schedules, and reporting of project status
■ Problem identification, recording, analysis, and resolution

According to the 12207 standard, the supplier is required to perform contract review activities and interfaces, and provide appropriate communication with the acquirer. Depending upon the requirements of the project, there are several possible reviews that could take place. Contingent upon the circumstances, these reviews can be formal or informal. The key to identifying the type of review is to determine who controls it. If the developing or servicing organization makes the presentation of the ICT, the review is considered informal. This type of examination is usually called a walkthrough. If the ICT is presented to a team or individual for examination prior to the actual review, it is formal and is usually called an inspection or audit.

Walkthroughs are very common in the ICT industry. They can be as informal as programmers demonstrating a segment of code and asking for help or regularly scheduled sessions that involve entire development teams or smaller working groups. The advantage of a walkthrough is that it requires little time, so the resource commitment and cost are minimal and negotiable within the contract. The disadvantage of a walkthrough is that it is not rigorous. The developers or service providers control the presentation, which means they can walk a reviewer past holes in their logic that no one will see or correct.

Inspections and audits have rigor, but they require resources and have substantial associated costs. The objective of an inspection review is that the developer or service provider surrenders the ICT for analysis by a third party (an auditor). The third party can be external or internal, but the analysis is essentially performed outside the development process, which leads to much higher rates of error detection and correction. In addition, inspections generate documentation in the form of reports and recommendations. Inspections and audits are almost always scheduled events and are often included within the project plan and contract.

In some cases, usually by stipulation of the contract, the supplier performs walkthroughs, inspections, and audits with representatives of the acquiring organization. The 12207 standard requires that any joint review activity must take place in accordance with the activities described in ICT Review process and Audit process. The supplier has the responsibility to perform the overall security and quality assurance activities in accordance with the standards assurance process requirements. Through this process, the supplier monitors and controls the progress and quality of the ICT products or services throughout the contracted life cycle.

Lastly, at the core of the inspection process lie the reports that are issued from the technical reviews. The supplier is obligated by the 12207 standard to report the results of all evaluations, reviews, audits, tests, and problem resolution meetings to the acquiring organization. Such obligations could be directed by specific language in the contract. The standard also requires that the supplier provide the acquirer access to the supplier's and subcontractors, facilities for review of ICT products or services as specified in the contract and project plans.

Supply Activity: Product/Service Delivery and Support

Upon completion of the development and testing process, the supplier will be ready to deliver the completed product or service to the acquiring organization. Recall from the discussions earlier in the chapter that the acquiring organization must be prepared to conduct a formal acceptance activity on the product or service. We mentioned earlier that the details of this activity are usually itemized in detail in the contract and acceptance plan. The plan defines the expected form of the final product or a detailed description of the service that the supplier is providing, as well as the methods for assuring all requirements have been satisfied. Most acceptance plans specify the test cases, test data, test procedures, and test environment that will be used for final assurance.

The acquiring organization carries the responsibility of defining and following the acceptance and operating process that will be used to accommodate the deliverable. These processes are developed directly from the acquisition plan, and they describe the specific steps required to ensure the security of the product or service. Among the activities of this process is a specific set of steps to ensure that basic security is built into the project itself. Security assurance in this context is essentially a process-based activity. The product is considered accepted when all conditions specified in the contract are satisfied. After acceptance, the acquiring organization implements the plan for how to deliver the product. The delivery process must be specified in the contract as well.

Once the product or service has been rendered, the supplier must certify that the product meets the terms and conditions of the contract. Customer acceptance is typically supported by extensive testing, reviews, and audits. Finally, the supplier will transition the product or service into the acquiring organization. To ensure a smooth transition between the two organizations, the standard also stipulates that

ISO/IEC 12207 Product/Service Delivery and Support Tasks	
6.1.2.3.5 Product/Service Delivery and Support Tasks	
6.1.2.3.5.1	The supplier delivers the product or service to the acquirer according to contract.
6.1.2.3.5.2	The supplier provides support of the delivered product or service, to the acquirer, according to contract.

Figure 2.14 Product/service delivery and support tasks.

the supplier should agree to provide routine technical support or installation service to the acquiring organization during the changeover.

Unless otherwise stipulated in the contract, the supplier is responsible for preparing the transition plan. This plan usually evolves directly from the requirements specified in the contract. The value of such a plan is that it allows the supplier to define the procedures to ensure that the product or service is transitioned in a safe and secure environment. Within the plan, the supplier defines the resources it will commit to execute the work as well as the scope and extent of customer involvement at each stage. Upon completion of the plan, the supplier installs the product in the customer organization or delivers the service according to contract. Figure 2.14 provides a summary of the two tasks of the product/service delivery and support activity.

Supply Activity: Closure

As each 12207 standard process has a defined activity that initiates activities performed during the ICT life cycle, each also contains an activity that draws a conclusion to a group of activities. That point holds true of the Supply process as well. The process ends when both parties acknowledge that all legal terms and conditions of the contract have been satisfied. The supplier accepts payment and then transfers responsibility for the product or service to the customer, as directed by the agreement. The terms and authorization for closing the project should be stated in the contract.

Chapter Summary

It should come as no surprise that successful SCRM requires that risk awareness and mitigation practices promoting legitimacy of information, identification, protection, detection, monitoring, and recovery mechanisms for security threats and vulnerabilities, and secure development processes be "built into" a well-defined life cycle process that is globally adopted by all parties throughout the supply chain.

The challenge is to get all organizations throughout the supply chain to agree on a common framework.

ISO/IEC 12207:2008 provides a defined set of activities and tasks within the standards Agreement Processes that aims to provide a single resource that all organizations participating in a supply chain relationship can adopt in order to provide a streamlined and risk aware life cycle process that stretches from the concept of need for a particular product or service through the contracting of a primary supplier, the primary supply interactions with subcontractors, to the verification, validation, and delivery of the product or service based on contracted requirements. The standard does so by addressing the activities and tasks performed by the customer (the acquirer) and an alternate set of activities and tasks performed by the supplier and their subcontractors.

The underpinning of the Acquisition process is to define those activities related to the customer role in purchasing system hardware, software, and services. The prescribed activities for acquisition are intended to convey the specific tasks that should be performed by any organization that has identified the need to acquire a hardware or software system or other form of a service that supports the larger scope of an ICT system. Further, the Supply process closely resembles project management activities. It coordinates and monitors the general function of providing a specified solution to a customer. It does not, however, embody development tasks. Instead, it is chiefly composed of various planning and control functions. This very generic view of the process legitimizes its implications to the underlying practices of SCRM.

Both the Acquisition process and Supply process are unique in terms of stipulation of activities and tasks performed by either the acquirer or supplier. However, there is a distinct element of dependency and collaboration between the parties, in many of the activities of both processes, demonstrating the need for common understanding of acquirer/supplier roles and responsibilities regarding interjecting risk management into the development process. Put differently, in order for SCRM to be effective, all parties must be on the same page in terms of the development life cycle process and the management of that process. The 12207 standard provides that mechanism.

Key Terms

acquisition plan: developed by the acquirer, this is a formal document that outlines the steps to be taken in order to deliver the predefined product or service

acquisition requirements: sometimes called a Request for Proposals (RFP), a formal notice issued by the acquirer to potential suppliers aimed at soliciting bids for the development of a required ICT product or service

concept of need: sometimes referred to as a feasibility analysis, identifies the problem domain and the justification for creation of a project to develop a solution

configuration management: a formal process utilized by the acquirer, supplier, and subcontractors to ensure the continuing status of a logically related array of system components; the detailed recording and updating of information related to the products and services that make up the acquiring organization's ICT system

contract: a legal document between the acquiring organization and its supplier, enforcing the agreement to deliver an ICT product or service

joint review: a project oversight activity that provides a mechanism for the acquirer and supplier to jointly perform verification and validation on parts of a product or service as it progresses through the development process or acceptance by the acquirer prior to delivery

requirements specification: a formal document containing all functional and nonfunctional requirements that will be necessary to provide a solution to the problem domain identified in the concept of need

security assurance: those processes that ensure the design and implementation of a system free of security threats and vulnerabilities, capable of mitigating future security risk

supplier oversight: activities performed by the supplier of a product or service that ensure that the development process adequately meets the contracted obligations. When subcontractors are used in development, supplier oversight extends to provide assurance in the work it is performing

validation: testing to ensure that the developed product or service provides the intended functionality at a high level of quality

verification: the process of testing documented product or service requirements against those specified in the requirements specification, in order to ensure that they have been met

Reference

International Standards Organization, *ISO/IEC 12207:2008 Systems and Software Engineering-Software Lifecycle Processes.* Standard, ISO, Geneva, Switzerland, 2008.

Chapter 3

The Three Building Blocks for Creating Communities of Trust

At the conclusion of this chapter, the reader will understand the following:

- The role and importance of communities of practice in supply chain risk management (SCRM)
- The relationship of communities of practice to supply chain trust
- The standard communities of practice and their characteristics
- The concerns and issues associated with each individual community of practice
- The general structure and principles of interaction among communities of practice
- Details of SCRM practice for each community of practice
- The role and importance of capability maturity in developing supply chain trust

Introduction to Product Trust

The constituent elements of a product supply chain are hard to identify let alone secure. This is a particularly difficult assignment where complex technology development and integration projects are involved due to the layers of complexity in the typical supply chain and given the fact that most products are integrated through a multilevel collection of suppliers who are often based offshore. Consequently, in this chapter, the reader will discover the importance of a formal, comprehensive, standards-based definition of the activities and tasks for the three common communities of practice in a supply chain. Process standardization is important because commonly understood and well-defined processes are critical to the overall assurance of trust.

Supply chain attacks are based on, and originate from, the commercial products and services that an organization acquires. Typically, these types of attacks might involve manipulating or corrupting an acquirer's hardware, software, or services at any point during their life cycle, from development to retirement. The commonly accepted method for addressing these attacks is based on an array of defense-in-breadth-and-depth controls, which the organization deploys as a means of protecting itself from known threats.

Most products, either developed or commercial off-the-shelf (COTS), are managed through multilayered, multivendor and even multicultural team approaches. All the activity at every level in the process must be fully coordinated and controlled up and down the development process to ensure trust in the eventual product. Coordination of multifaceted elements of work requires a commonly recognized and accepted conceptual model of coherent control processes and control activities. The aim of that model is to fully encapsulate and relate the processes and activities a given set of managers will utilize to understand the precise security status of any given product as it moves up a supply chain from initial design to final product integration, testing, and assurance. The specific purpose of this chapter is to give you a working understanding of the three communities of practice that must be considered when developing a defense-in-breadth-and-depth solution. These are the acquirer, supplier, and integrator communities (Figure 3.1).

The ever-increasing reliance on globally sourced hardware and software components leverages the growing importance of the assurance of product trust. The accelerating trend toward multinational approaches to creation and integration of COTS type products makes it almost impossible to utilize a given supplier's reputation or even overall corporate ownership as a basis for assurance. This is due to the construction of complex things involving a product tree that integrates parts from

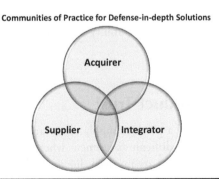

Communities of Practice for Defense-in-depth Solutions

Acquirer

Supplier   Integrator

Figure 3.1   Three communities of practice for defense-in-depth solutions.

a diverse range of small suppliers into products of ever-increasing size and scope. This tendency will obviously reduce any given organization's ability to understand and track the various components that make up a final product.

Globalization is an enduring fact of life in our era and is likely to have increasing influence over time; thus, all organizations need a standard capability to assess and manage supply chain risks. This capability is necessary to ensure a trusted supplier base, and most large organizations have relationships with a wide range of suppliers that directly or indirectly provide software, hardware, services, or human resources that are involved in a range of critical information processing functions in an organization. The problem is that failures in the supplier community can cause weaknesses in the product that are beyond the control of the customer organization. These weaknesses can be exploited, so the need to identify, assess, and specifically mitigate information and communication technology product supply chain defects is crucial to overall product trust. Mitigation is done by implementing a relevant set of electronic and management controls.

Typically, customer–supplier relationships are specified in a legally enforceable contract. Both the supplier and acquirer share the responsibility for defining and achieving the security goals embedded in that contract. Given the requirements for stable and continuous coordination of the customer–supplier relationship, a fundamental set of commonly adopted processes needs to be planned and implemented in order to uphold that relationship. This includes the deployment of all necessary governance, business management, operational and human resources, and management controls to establish robust protection against attacks on the supply chain, whether these threats are intentional or accidental.

Building a Basis for Trust

Most modern information technology operations are built around sourced products or services. The suppliers of those products and services provide critical components of the customer operation including software, hardware, processes, or human resources. The purchasing process must establish a defined set of associations between the acquiring organization and a given set of suppliers. Nevertheless, the acquiring organization will most likely not be able to monitor or at least directly control the production and delivery processes for the products they are purchasing. Therefore, potential security risks lurk in every relationship between acquirer and supplier base. A requisite level of trust between the acquirer and all its supplier organizations should be established and assured by the definition and implementation of a formally established, commonly accepted, security governance process— one that contains a well-defined set of auditable and effective controls.

In all customer–supplier relationships, both entities need to take responsibility for the establishment of a prescribed, joint-coordination process, one that will adequately address all the known trust issues that might be present in each product

Acquisition process, to ensure that effective management, operational, and human resources management practices are embedded in the customer–supplier relationship. The attendant best practices must be planned and implemented in such a way that they provide an effective day-to-day process. This highly integrated set of practices should provide direct support for the responsibility to ensure that the sourced product is kept under strictly managed control.

The Hierarchy of Sourced Products

Instilling trust in sourced products can be approached by requiring the supplier to obtain an independent credential proving its capability, such as ISO 27000 certification or FISMA approvals to operate. Both standard frameworks implement and document a commonly recognized set of auditable best practices for the overall supply chain process. The form and application of those assurance certifications obviously need to be agreed on by the acquirer and supplier as part of the selection and contracting process.

Trust becomes a much more complex issue where proof of supply chain security is required, particularly with global supply chains. An information and communication technology product supply chain includes those organizations linked in a common effort to develop a product or provide a given service to customers. An information and communication technology product supply chain usually involves a multitiered set of successive contractually defined supplier relationships, which are normally arrayed in a one-to-many hierarchy. The end product or service typically comprises coherent components or modules, plus any related resources and processes that are produced, whole or in part, by another supplier. The supplier integrates these components or processes into a single distinct entity at their level in the supply chain. Because of this hierarchical arrangement, the eventual information and communication technology product or service is likely to have been sourced and created by multiple suppliers.

This produces multiple roles and responsibilities for any organization that is operating within an information and communication technology product supply chain. This happens because every supplier is also an acquirer in relation to the upstream elements of the process, while the same organization is a supplier in relation to the downstream organizations.

From the perspective of the organization providing the product or service, the downstream organization is termed the customer, whereas the customer at the eventual end of the information and communication technology product supply chain is referred to as a consumer or the acquirer. Generally, the end customer or acquirer has limited control over their direct supplier's security practices and the end customer has no control whatsoever over the security practices of any organization beyond the direct supplier within the supply chain. Therefore, the acquirers at the downstream level and suppliers at the upstream level in the information and communication technology product supply chain need to manage the common risks that crop up, as the eventual product moves down the supply chain. Risk management can be a challenging task because each component in the supply chain is a distinctively different entity. Due to differences, the security risks that are associated with individual supplier relationships in the supply chain can represent a serious concern, not only for the customer organization, but also for the consumers of the product or service and all other parties involved in the process. Therefore, effective supply chains are built around fundamental trust.

To ensure the proper level of trust, acquirers and suppliers must be equally responsible for maintaining the integrity of the contractual agreements and for managing all known supply chain security risks, including establishing the precise definition and agreement on operational security roles, responsibilities, and functions within the SCRM process, as well as the mutual implementation of a tangible set of appropriately designed and agreed-upon controls.

In the case of the supply chain itself, each inherent customer/supplier relationship is designed to accomplish a specific set of business goals. The number of such relationships is likely to grow with a major product or service, with the creation of a supply chain, and with a couple of customer/supplier levels on the outer boundaries of expansion, which is conventionally set at five levels. Figure 3.2 shows the upstream supply chain relationships to the left of the acquirer organization and the downstream relationship of the customers to the right. Visibility and control are hard to maintain at any level further down the chain than the direct, e.g., subcontractor, level. This lack of control is likely to result in relationships being badly managed or even unmanaged by the customer.

Even worse, a complex organization is likely to contain a wide range of different internal entities in its routine operation, all of which are likely to maintain a totally unrelated set of supplier relationships. Given the potential number and

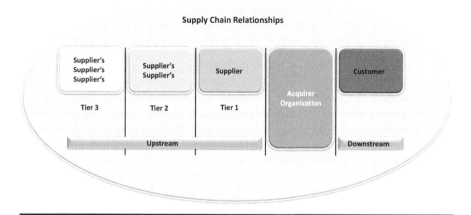

Figure 3.2 Supply Chain Relationships.

diversity of those relationships, control is likely to be established using a wide range of different practices. This fragmented approach to supply chain relationship management becomes even more problematic when each of the individual supply chains includes multiple levels. Thus, the diversity of supply chain activity will almost certainly guarantee that the organization will not be able to suitably address all the risks to its products and services, within all its necessary levels and relationships.

This problem can only be appropriately addressed by formulating a legally enforceable contract that defines the roles and accountabilities of all of the dispersed acquirer operational elements involved in the supply chain. The agreement must state the mutually acceptable common controls and responsibilities for implementing a SCRM process among all parties in the supply chain. A formal, legally enforceable agreement is critical in the assurance process because the lack of clear coordination and control of the deployment of security measures is likely to impact the supply integrity chain in the following ways (NIST, 2016):

■ Misunderstandings between acquirer and supplier, or differences in social, cultural, or organizational operational control behaviors, can lead to gaps in the actual security governance, risk tolerance, and compliance practices within a supply chain.

■ Implementation of unintended control behaviors or the injection of unknown or unplanned risk dependencies due to the misunderstanding of the various organizations' responsibilities for risk mitigation/risk control within the supply chain can lead to exploitable gaps in the defense-in-depth solution.

■ Conflicting or dysfunctional acquirer and supplier information security controls will serve to weaken the overall end-to-end supply chain risk identification and management capabilities of both parties. For instance, weaknesses, mismatches, or conflicts in the governance framework, loss of security control by higher-level entities in the hierarchy, or conversely, unanticipated, unknown, or undesired outsourcing relationships between lower levels in the hierarchy will lead to breakdowns in the process.

■ Miscommunicated or misunderstood control practices that might be implemented by the supplier but do not satisfy the risk requirements, risk tolerances, or risk appetite of the acquirer might make the acquirer vulnerable to already identified risks, that is, if those risks have been presumed to be addressed and mitigated by the supplier. For instance, business continuity, or disaster recovery planning might be absent, ineffective, or improperly executed by the acquirer because it was assumed that the supplier had that issue covered. Of course, these kinds of interruptions can lead to unavailability up and down the supply chain (Figure 3.3).

These issues all point to the need for coordinated and intelligent management of the actions of all participants in the supply chain, which implies the need for a formally established and documented SCRM scheme. In this scheme, the acquirer

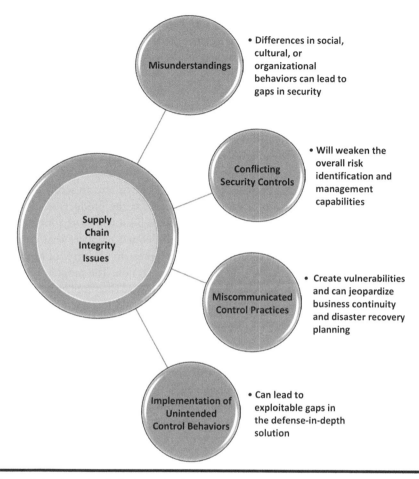

Figure 3.3 Supply chain integrity issues.

and supplier must evaluate risks and then select, implement, and maintain the requisite controls to mitigate them. In the context of supplier relationships, the typical control categories consist of those control functions and behaviors that directly address supply chain security risks.

Those controls are normally associated with assurance steps that are taken to mitigate meaningful risks in an upstream supplier's product or service. The controls themselves typically assure or control identified risks to supplier products, which can impact the supply chain security relationship further up the chain. Controls also enforce proper understanding of the operation of the overall execution of the process by enforcing visibility into the assurance operations of the other organization. That includes enforcing control of routine management reporting, as well as the monitoring, auditing, and certification of product and service correctness. The need for certification and accreditation is normally itemized in the contract. In simple terms, the contract between the acquirer and the supplier must commit both organizations to the implementation and maintenance of an agreed-upon set of controls.

The issue of transparency in the assurance of the supply chain's functioning is perhaps the most critical factor in the entire assurance process. Trust between any given set of entities depends on the ability to clearly see and understand the internal workings of the relationship on both sides. Therefore, documented and auditable assurance that the supplier has established adequate SCRM is an essential part of ensuring a long-term trusted relationship. In most cases, to do this properly, the acquirer needs to evaluate the adequacy of the supplier's product or service controls based on a mutually agreed-upon set of criteria for supply chain security management. The criteria must mirror what the acquirer deems adequate to mitigate risks to an acceptable level.

The acquirer's acceptance of a supplier's process for the production, delivery, and operation of the products and services that have been contracted for delivery should be based on commonly accepted and well-defined practices that ensure a sufficient level of supply chain security. The level represents the level of assurance the acquirer wishes to maintain within its own organization. The supplier is judged based on what the acquirer considers acceptable, not what the supplier is currently doing to address supply chain risk. The criteria for making this evaluation should include the itemization of explicit practices for the direct management of

- Every known physical and electronic threat to the prospective product or service
- The integrity of all electronic and other components in the supply chain
- The basic integrity of supplied components
- Legal, regulatory, and environmental supply chain security risks that may impact the acquirer's supply chain security within a given local situation
- The physical integrity and correctness of supplier facilities

This assurance applies to every aspect of the suppliers' business as it relates to all downstream suppliers, as well as interaction with upstream acquirers. The actions required normally represent a complex set of items; to appropriately address supply chain security throughout the information and communication technology product supply chain, all participants in the process need to agree on and deploy a standard, common, coordinating framework as a point of reference.

The following set of standardized organization-wide processes for the acquisition of products and services should be represented within this framework:

- The ability to assess and monitor all security risks associated with the supply chain
- The ability to ensure any compliance requirements for the secure exchange or sharing of SCRM information
- Continuous monitoring of individual supplier performance within the supply chain, especially as it applies to any alterations in the relationship between suppliers

When COTS elements and services are part of the Acquisition process, it is also essential that the acquirer require formal documentation that demonstrates standard, good design and development principles and practices. The implementation of effective SCRM begins with the fundamental performance of good system design and development practices (Figure 3.4). Examples of good practices are (NIST, 2016)

- Upward integration of supply chain security requirements from inception of the project
- Assurance that the supply chain security, acquisition personnel, legal counsel, and other appropriate advisors and stakeholders are participating in all risk management decisions
- Assurance that the resources allocated for supply chain security and SCRM are both adequate and effectively disbursed.
- Adoption of well-defined, properly documented, and repeatable processes for the specification of product and service contracts
- Design, development, and implementation of effective contingency plans for continuity management
- Defined processes for adequate oversight of suppliers, including Service-Level Agreements (SLAs)

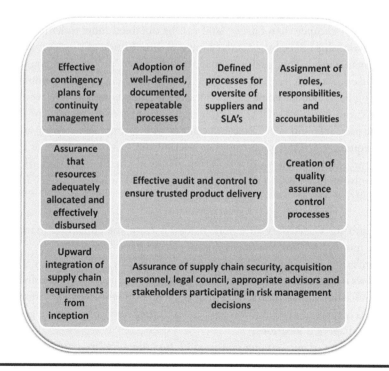

Figure 3.4   Examples of effective SCRM.

■ Effective audit and control to ensure trusted product delivery, including a full set of performance criteria to be used as a basis for judgment, vetting, and acceptance
■ Creation of an active quality assurance and quality control process
■ Explicit assignment of roles, responsibilities, and accountabilities so that all participants will know what actions to take and who will take them

Organizations should utilize a fully defined and integrated set of processes and activities to consistently and effectively assess and mitigate supply chain risk. Therefore, the first step in the attempt to build effective supply chain trust is the development of a strategic set of organization-wide plans and directions. These plans and directions must clearly prescribe and establish the complete set of roles and responsibilities for all constituencies involved in the SCRM process. In this respect, the organization must instantiate a well-defined set of security best practices within a unifying meta-framework of related SCRM process.

The Problem with Sourced Products

Sourced products offer a huge business advantage over custom developed systems, the same advantage buying a suit off the rack has over bespoke tailoring. COTS products are much cheaper than custom and can be obtained more quickly. Therefore, COTS relationships have become the source of a multitude of products or services for any organization. This includes outsourced IT work, professional consulting or other kinds of services, renting or leasing basic equipment such as computers and networking gear, cloud computing services, research and development work, classic communications utilities such as backbone, and other types of Internet connectivity.

These products and services are obtained from a supplier or group of suppliers; consequently, the customer/supplier relationship has become one of the most integral and fundamental routine building blocks of any IT operation. The extent of the involvement with COTS will vary for a range of business reasons, including (NIST, 2016)

■ Acquiring a necessary function or service from the consumer market saves the money and time spent developing the product from scratch.
■ Purchasing short-term or highly specialized competency from the marketplace allows the organization to immediately address or achieve specific business goals.
■ Purchasing rather than developing new functionality lets the organization quickly react to market forces such as setting up new operations in geographically dispersed locations.
■ Purchasing or outsourcing marginal IT functionality from an array of suppliers and contractors lets the organization prioritize its resource investment and ensure that essential core operations are always maintained at optimum levels.

A process of managing risk to sourced products normally calls on a single organizational entity for the design, development, and implementation of the process. The purpose of that central entity is to coordinate the basic definition, description, implementation, and resourcing requirements of the practical SCRM response. The creation of that entity can be ad hoc or formally documented. The aim of this aspect of the process is to design and then enable a set of useful practical best practices, which can accurately identify all relevant supply chain threats, vulnerabilities, and weaknesses and then execute a comprehensive risk analysis to describe the operational parts of the supply chain that might be affected. The product of this strategic risk-identification/risk-analysis process is an effective, real-world, practical approach to ensuring overall trust in any given acquisition sourced through the organization's supply chains.

A customer forms a relationship with a supplier to purchase a given product or service. This intention is normally fleshed out in a mutually agreed-upon specification of functional and nonfunctional requirements. The relationship typically lasts a predetermined legally enforceable contracting period. The common hazards that might be present during that period were outlined in the first chapter; however, to summarize, those are

- *Malware* inserted into the product up and down the supply chain during its development
- *Counterfeits* sold as part of a system but do not meet specified requirements, although they might appear to meet functional needs
- *Breakdowns in production* leading to the operational unavailability of the system or service as needed
- *Supplier insufficiencies in performance* leading to inadequate or faulty products or services as delivered
- *Latent defects in the product* leading to failure or adversarial exploitation

Three of these concerns are directly shaped by failures in the supplier's production process: malware, counterfeits, and defects. They stem from the inability of both parties to maintain adequate oversight and control over the threats associated with production. Consequently, the two parties need a formal monitoring or auditing process to oversee the production process. This is a joint effort designed to underwrite trust in the ultimate product. The other two concerns, supplier breakdowns and supplier insufficiency, are a consequence of the initial supplier selection. This selection process normally involves vetting/reputational issues.

Thus, we focus on the means of developing a standardized range of jointly agreeable controls that can be deployed to facilitate effective risk management among the communities of practice involved in the SCRM process. Obviously, the lack of an effective comprehensive and in-depth risk management process among the participants in the product or service procurement process can adversely impact the customer's ability to trust the security and integrity of the COTS product itself.

The lack of effective and meaningful risk management practices will have an impact on SCRM in the following generic ways (NIST, 2016):

- Miscommunication and misunderstanding among the participants in a supply chain can cause the involved parties to adopt approaches that do not effectively monitor or accurately ensure the integration of the acquirer's security requirements with the supplier's assurance approach. Besides the potential for malicious activities, this breakdown in communication can lead to failures and disruptions in the delivery of the product.
- The existence of gaps in security practices and controls between acquirer and supplier can cause critical failures in the enforcement of security at some point in the supply chain. This might be in the form of failures in the threat identification and mitigation processes or the formal security compliance practices of the involved parties.
- Finally, and worst of all, it is possible that the approach to security practice will so greatly conflict among the members of a supply chain that it will impede or weaken the security of the overall process or the product itself.

Several categories of risk management must be considered during the life cycle of a supply chain. First and foremost, there is the issue of overall management governance and control. It should be self-evident that a lack of formal, coordinated management or a weakness in the formally deployed control set will cause the participants in a supply chain to lose coordinated command and control over the actual nuts and bolts of the process.

From a threat standpoint, and on par with the loss of effective governance control, is the issue of unregulated outsourcing. In that respect, without the presence of organized outsourcing control practices, it is possible that suppliers could subcontract a subset of the product or service contract to another, more untrustworthy supplier organization without the acquiring organization's knowledge. This situation will inevitably reduce the customer's visibility and control over its purchased products and potentially open the customer organization to additional unknown hazards. Outsourcing risk includes joint operational issues such as the possibility that the controls that have been established by the subcontractor will not address the risks the acquirer considers meaningful.

Another obvious consequence of a failure to establish formal coordination and alignment in the risk management process up and down the supply chain is the potential for miscommunication and misunderstanding between the various parties involved in the development process. Having the ball fall between the outfielders because of badly defined communication practices will leave the customer organization vulnerable to threats that it believed had been identified and mitigated by the supplier.

Furthermore, if the communication process is faulty or missing, the acquiring organization's specific requirements for confidentiality, integrity, and availability may not be adequately communicated to the suppler. Thus, crucial information security requirements may not be addressed adequately. In addition, if important information about delays or breakdowns in service is not available up the supply chain to all parties during the delivery of the product or service, the response to the delay might not be timely. Failures in communicating breakdowns at lower levels in the supply chain might cause service interruptions at all levels.

Finally, if there is a breakdown in communication between the acquirer and supplier organizations, suppliers might fail to allocate sufficient human resources including appropriately skilled staff to address the additional work required to solve the immediate problem. Lack of proper communication among the participants in a supply chain accounts for 40% of the categories of supply chain risk cited by the Office of Management and Budget (OMB)—specifically breakdowns in production and supplier inadequacy. Therefore, an effective communication process is an important fundamental condition in maintaining a proper trusted product assurance process.

Finally, with respect to the generic issue of difference in organizational cultures, without strict understanding of the terms and conditions of project work between partners, it is likely that the development or Acquisition process itself will be conducted dysfunctionally. Even worse, elements of the product or service responsibility will violate the customer's expectations regarding product security and integrity. The issue of geographical, social, or cultural misunderstanding is the direct result of an insufficient grasp of local cultural requirements, which exist at all levels in the supply chain. A failure in communication plays directly into the issue of geographical, social, or cultural misunderstanding. For example, a supplier who is not familiar with local regulatory requirements might inadvertently violate a law or regulation, which could lead to financial penalties or reputational damage for the customer. On the other hand, the misunderstanding of a contract stipulation or the conditions of a standard requirement could cause a legal dispute between the two parties in a supply chain relationship.

Implementation of a SCRM process normally requires a single organizational entity to do the design, development, and implementation of the process. The creation of that entity can be ad hoc or formally documented. The aim of this aspect of the process is to design and then enable a set of useful practical best practices, which can accurately identify all relevant supply chain threats, vulnerabilities, and weaknesses and then execute a comprehensive risk analysis to describe the operational parts of the supply chain that might be affected. The end product of this strategic risk-identification/risk-analysis process is an effective real-world, practical approach to ensuring overall trust in any given acquisition sourced through the organization's supply chains.

In this respect then, the organizations should adopt a standard, organizationally accepted SCRM model to guide the creation of a documented and adequately understood set of coordinating processes that sufficiently itemize those roles and responsibilities. The outcome must be suitably detailed to guide the conduct of supply chain risk mitigation activities. A common set of real-world procedures has to be documented, approved, and promulgated to the organization as a whole, in order to ensure that the SCRM process is executed in a universal and uniform way. The procedures must assign and fully describe the participants in risk assessment, risk analysis, and risk decision-making for the organization.

Besides the persons responsible for the actual supply chain risk mitigation process, the supply chain stakeholders should be identified and involved in the process within their individual areas of authority. Supply chain risks are both pervasive and dispersed, so they need to be addressed by every stakeholder who has skin in the game and might provide a useful perspective. This includes everybody from the actual executive owner of the supply chain all the way through security personnel within a given sphere of operation, the stakeholders who might be involved with the products and services that are associated with a given supply chain product, and possibly third parties who might provide additional perspectives.

Promoting Trust through Best Practice

The following is a set of standard practices intended to promote trust in the acquisition, development, and operation of the COTS Acquisition process. The aim of these practices is to ensure the correctness and integrity of stated cost, schedule, and performance requirements up and down a worldwide supply chain. The practices themselves are designed to underwrite an effective risk management program the average customer organization can implement as standard operating procedures in its supply chain.

Logically, an organization needs to adopt a methodology for handling supply chain risk in order to ensure that the requisite set of controls is properly embedded in the procurement process. That methodology should address four global questions (NIST, 2016) (Figure 3.5):

1. How will supply chain management criteria be developed and enforced?
2. What are each party's roles and responsibilities for secure product acquisition?
3. How are supply chain risks factored into day-to-day product acquisitions?
4. What are the controls necessary to eliminate all reasonable supply chain risks?

From a practical standpoint, the solution is characterized in the everyday best practices of the organization. Thus, the first step in the effort to embody comprehensive SCRM lies in the implementation of a well-defined set of standard and systematic best practice controls. Ultimately, information security risks lurk in all of the nooks and crannies of the acquirer/supplier/integrator relationship. Therefore, to assure a trusted product, the practices that will govern the three areas of supply chain practice must be fully and appropriately defined and implemented.

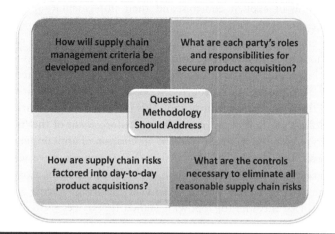

Figure 3.5   Questions methodology should address.

Much of this requirement centers on the need to ensure that all three of the parties in a supply chain know and fully understand each other's practices and means of doing business. Justifiably, that understanding depends on the free flow of information during the actual development of the product and afterwards in the form of detailed documentation. This information must be appropriately and correctly defined, developed, and protected by means of a standard Agreement process. That process was itemized in Chapter 2 by the Agreement processes of ISO 12207:2008; it will be discussed throughout this chapter.

Moving the Product up the Supply Chain

Supply chains represent a delivery path defined by a set of interlocking contracts. The aim of this delivery path is to enhance visibility in the evolution of the product. If the steps in the supply chain are well defined in advance, it is much more difficult for an adversary to sabotage or subvert a product as it is developed. This path is built around a minimum set of baseline practices for supply chain delivery. Logically, the path is designed so that the delivery of the evolving product is subject to the minimum amount of exposure to or access by unknown or unvetted sources.

The product is normally a contracted system delivered to a given organization in the supply chain. As supply chains are dynamic and therefore subject to subversion, contractual requirements must include a stipulated set of steps to protect, monitor, and audit the product's elements as they move up the supply chain. Therefore, the assurance of personnel, particularly those directly involved in the supply chain process, requires that specific controls be defined and implemented to protect the movement of operational components up the supply chain.

The definition of explicit controls and their implementation is particularly effective in the case of multitiered supply chains, as well as any form of outsourced development environment. This generally requires the organization to leverage its existing methods of control to create continuous monitoring activities to ensure the evolution of the product up the supply chain. This is normally specified in the overall contract for the product or outsourced services. The contract defines the means of protecting the product from subversion. This is done through a variety of methods such as robust configuration management, enforcement of the first principles of security, cryptographic methods, and the application of antitamper techniques.

As we have already seen, the process itself is supported by a wide range of complementary monitoring and audit approaches. The purpose of those approaches is to leverage data gathered into a means of ensuring against supply chain vulnerabilities. That includes the following standard practices (NISTIR 800-181):

- Monitor and audit supplier system activities for compliance with requirements, as well as detect commonly known supply chain concerns
- Monitor and audit supplier system activities to detect the presence of malicious functionality, known vulnerabilities, or changes in suppliers or supplier personnel
- Monitor and audit systems and operations to reduce the risk of unauthorized element(s) removal, replacement, and/or modification

The discovery of a supply chain vulnerability requires all supply chain stakeholders to initiate a formal vulnerability management process aimed at mitigating any adverse consequences. This requires the establishment of a process for managing supply chain vulnerabilities, including detecting, tracking/logging, selecting a response, performing the response, and documenting the response. Incident attributes that might be reported and recorded include the time and place of the anomalous behavior or discovery, the method or mechanism by which the incident was detected, and the suspected or known identity or source of the incident. Formal vulnerability management is initiated when a flaw, defect, or adverse action is reported and recorded.

The first step is to determine the potential impact of any exploitation and includes tracking any identified dependencies across the supply chain. Dependencies are tracked and analyzed, whether the vulnerability is being exploited or not; this includes all data about the form, actions, and impacts of the vulnerability, if known. It also includes specifications of where the vulnerability was introduced in the supply chain. The level of risk that this specific vulnerability represents is determined, and a decision is made to mitigate or accept the risk. Risks are prioritized by critical elements first. This is a normal defense-in-depth arrangement where the decision is based on the potential for higher severity of consequences.

Once the decision is made to address the problem, the organization takes the necessary steps to remediate it as well as to identify the causes associated with that incident. The specific aim of the process is to identify a root cause. This is done to both determine the likelihood of damaging exploitation and eliminate the weakness that originally produced the problem. If a supply chain incident occurs because of a newfound vulnerability, then the organization must have a process ready for managing supply chain incidents. The process includes implementing a well-defined response to any potential incident that might have been identified in advance as well as criteria for deciding what would constitute a supply chain "incident" for any given component, process, or product. Criteria might include scope of incident occurrence, the classification of incident types, and a list of appropriate responses. Then the stakeholders in the supply chain collaborate to ensure appropriate resolution for all affected parties. This requires the organization that made the discovery to coordinate its supply chain incident management activities with all other stakeholders to ensure consistent and effective management of the incident.

Normally the affected parties will define and assign specific individuals to collect, process, and disseminate SCRM incident information; develop and maintain the timeline and method for incident monitoring and reporting; and collaborate with all relevant stakeholders to ensure appropriate resolution for all parties. This requires that the responsible parties implement an appropriately defined supply chain incident management process and coordinate all relevant supply chain incident management activities with other organizations to ensure consistent and effective SCRM.

The SCRM team will normally perform some form of forensic analysis to determine the cause of failure. The team will also assess the impact of the failure, as well as the necessary steps to mitigate its actions. Finally, it will document and report the incident and all related findings, as defined in the plan.

The Standard Approach to Identifying and Controlling Risk

The literature is full of mechanisms for identifying, collecting, and labeling potential risks: methodologies to retrospectively test and review results, assess project performance, and obtain insights from customers and stakeholders; static test processes market analyses; and even blue-sky thinking. Once the relevant risks have been identified, the people responsible for the associated supply chain need to involve the affected parties—managers, policy advisors, and decision-makers. The technical leads for the product might be consulted along with the various program managers and perhaps even the organization's legal counsel.

The organization includes all potential sources of knowledge about potential risks to ensure that it has gotten a realistic and comprehensive understanding of those risks and ways they can be managed. The end result of that consultation process is a complete and correct identification of the risks involved, as well as the deployment of a set of applicable controls for managing the risks within a given supply chain. These controls are normally formulated into an operational baseline guided by a commonly accepted set of best practice controls.

At least three commonly recognized comprehensive security control frameworks can provide a specification of standard best practices for SCRM. These are the ISO 27000 International Standard, the U.S. Government's NIST SP 800-53 Control Model, and the Information Systems Audit and Control Association's (ISACA) Control Objectives for IT (COBIT5). Any of these represents a sound foundation or point of reference for the determination of the right set of standard security controls for an individual organization within a supply chain. Naturally, the standard controls specified in the model itself must be tailored or customized to the specific organizational application they are meant to address. However, for control and coordination's sake, the control set itself has to be consistent up and down the supply chain in order to properly address the specific threats to that particular supply chain.

When organizations determine the information technology requirements of a new or modified business security response, they should consider the specific protections against supply chain threats by employing an organization-defined list of measures as part of a comprehensive, defense-in-breadth strategy. Each practice represents a blend of programmatic activities, validation/verification functions, and general and technical requirements for every product. In many cases, the practice will apply to a software supplier and a hardware supplier separately, since most hardware devices contain some level of firmware or software that is integrated into the product.

The Three Standard Supply Chain Roles

The term "supplier" is used to describe the organization that produces a specific element and then provides it to the integrator for integration into the overall system. The supplier is synonymous with vendor and manufacturer. The term "integrator" generally describes a third-party organization that combines the product elements produced at a given level in the supply chain and thereby produces larger system elements at the next level in the build. The term "customer" is used to describe the organization that acquires or purchases and eventually operates a specific product composed of program elements that have been integrated as part of the supply chain development process. The term "acquirer" is synonymous with customer.

The requisite practices of all three are combined into a single prospective risk management approach. A broad range of potential SCRM approaches can be applied to an information system or elements of an information system. Nonetheless, that application will vary depending on whether the actual organization is doing acquisition, supply, or integration work. This is an important distinction because the responses of an acquirer, integrator, or supplier require requisite best practices that by necessity diverge widely, exacerbated by the relative position of that organization in the supply chain. Thus, the organization's management and information security professionals need to select a given set of standard specified practices sufficient to mitigate supply chain risks in each situation.

The Acquirer Role

The acquirer role is responsible for the design and documentation of a detailed, formal, end-to-end, product sourcing/Acquisition process. The design should embody the form and structure of the Agreement processes described in ISO 12207:2008, including complete specifications of the preferred operational practices, acquisition strategies, and procurement activities to be employed up and down the supply chain. The acquirer's primary mechanism for enforcing this design is the contract, and transparency up and down the supply chain is indispensable for ensuring that all stipulated terms and conditions have been adequately addressed. The aim of the acquirer is to maintain optimum visibility for every aspect of the processes the integrator and supplier employ in the development and deployment of the product or service. The key objective is to obtain and maintain systematic knowledge of the activities of all participating organizations in the supporting tier of the development and integration process (e.g., the organizations further down the supply chain). Transparency is essential because it allows the acquirer to understand how the components of the purchased product or service are selected, created, tested, delivered, supported, and protected throughout their life cycles. To initiate that process, it is advisable for the acquirer to develop a procurement process that aligns with national, governmental, or international standards for process and product correctness. The aim is to introduce uniformity into the procurement process. If a component is developed using the same methods and processes as those used by another supplier, then it can be assumed that both components are uniformly similar.

Achieving the necessary uniformity might require the acquirer to promulgate policy and procedures that include comprehensive supply chain risk status assessment as a precondition for supplier selection and development of the product integrator's supply chain. Such standard assessments should identify and remediate each supplier/integrator's points of vulnerability using commonly recognized attack patterns, tools, methods, and procedures for that product. The goal is to strike the most effective balance possible between cost and risks given a known set of adversarial tactics, techniques, and procedures. Specifically, acquirers use information about risky integrators and suppliers and SCRM practices to identify and judge the necessary trade-offs among the desired level of product performance, the level of acceptable risk, and the resources that must be arrayed to maintain a proper and effective balance (Figure 3.6).

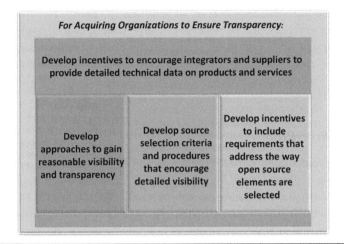

Figure 3.6 To ensure transparency, acquiring organizations need to.

For the acquiring organization to ensure sufficient and persistent transparency, it must (NIST, 2016)

- Develop incentives that encourage integrators and suppliers to provide program-specific detailed technical information and technical data on products and services throughout their life cycles
- Develop approaches that encourage integrators and suppliers to gain reasonable visibility and transparency into their ancillary supply chains
- Develop source selection criteria and procedures that encourage integrators and suppliers to provide detailed visibility into elements, services, and processes as part of their submissions for contracts
- Develop incentives that encourage integrators and suppliers to include requirements that address the way open-source elements are selected

The aim of these actions is to encourage product integrators and suppliers to provide accurate, up-to-date information about the functioning of their supply chains. This is meant to continue for the life span of the contract. The emphasis is on life cycle identification and management of risk to the system and its constituent elements. This includes the specification of all individual criteria and requirements for acceptance of all open-source elements integrated into the product build.

The acquirer should incentivize this process by seeking integrators who will provide necessary technical details about their products and/or services, including important documents like architectural drawings, wiring schematics, and/or entry and exit points for interfaces. Such information may also be important to long-term maintenance and support of the product should the integrator stop supplying the system/service element.

Integrators need to provide the capability to understand, evaluate, and document any aspect of their system elements or development process that could result in an exploitable weakness or vulnerability. The contract should specify explicit criteria for evaluation of security risks; this specification needs to state suitable tests that would be used to judge acceptable security, including routine data to be gathered, detailed types of activities, and measurements results. This evidence must demonstrate standard agreement with all contractual terms and conditions.

Suppliers need to be able to provide proof of comprehensive compliance with best practice in SCRM. This documentation needs to apply to every aspect of the delivered product, virtual and physical, as well as all integrated products from outside vendors including open-source elements. Since the essence of this proof lies in the processes that each individual supplier organization has adopted, it is particularly essential to maintain full transparency regarding the activities and tasks those organizations regularly perform, with emphasis on any process that might cause vulnerability or risk. It is also necessary to document and rectify any activities that deviate from contractual specifications in such a way that they might represent risk and subsequently follow up on reported anomalies to determine that any risks have been mitigated.

In support of this, the acquiring organization needs to develop and put in place a formal means to assess all relevant supply chain characteristics that include documenting every aspect of the physical and logical elements of the product. The aim is to provide auditable evidence of a set of security practices within the supply chain that will serve to satisfy all contractual requirements. Besides assurance of the supply chain, the acquirer should develop a detailed set of acceptance testing procedures that will validate COTS products as standalone elements or as integrated into a larger system. That includes identification of all subcontractors up and down the supply chain. The aim is to develop a full array of acceptance testing measures that will help the acquirer to better identify and evaluate defects that might lead to the compromise of products, processes, or services within the supply chain.

All supply chain activity must be controlled; thus, the acquirer needs to design and implement a baseline of tangible security controls. These controls are implemented as a system by all three participants in the normal supply chain: acquirer, supplier, and integrator. The controls are put in place to mitigate supply chain risk. The implementation process is underwritten by a formal risk assessment and analysis of all management, operational, and technical elements of the product. The purpose of this assessment is to ensure that all pertinent elements, processes, requirements, and business practices up and down the supply chain operate in a harmonious way that protects against compromise.

To do this, the acquirer must work with all relevant stakeholders to clarify all known risks and threats to the proposed product and then develop and promulgate a standard acquisition strategy that mirrors the recommendations of the ISO 12207:2008 Agreement processes discussed in Chapter 2. The specific process and activity requirements for the protection of the supply chain need to be identified and spelled out in the form of detailed work instructions. This will normally lead to the incorporation of an explicit set of management, operational, and technical controls whose purpose is to insure the supply chain against an identified set of risks. The acquirer will normally work with the stakeholders to understand, clarify, and prioritize all the risks and threats to the organization into a practical defense-in-depth scheme.

In many cases, the supplier is also an acquirer; it is necessary for each supplier to develop and put in place a means to identify and document the characteristics of its physical and logical supply chains. This is necessary to identify and mitigate any weaknesses that could result in the compromise of supply chain processes, including the ability to confirm that supply chain components have the capability to determine and report activities that could lead to the exploitation of a product element. This includes the ability to assess and confirm the effectiveness of supplier/integrator risk management policies and procedures throughout the product life cycle.

To make this effective, the supplier needs to adopt methods that can look across individual organizational boundaries within a given supply chain to confirm that the successful fulfillment of contract requirements does not lead to unacceptable risk. The latter requirement implies that the acquirer needs to develop and deploy acquisition policies and procedures in accordance with good defensive design practice. The purpose would be to make the supply chain less visible and predictable to any potential adversary who might be bent on subverting it. Thus, the bidding and acceptance process needs to consider limiting the product channels by which the element is acquired in a way that makes it hard for an adversary to determine when and where an element will be acquired. Because the product can be subverted in transit, it is also important to limit information about the actual delivery process and testing methods. Usually, dynamic sourcing is adopted to make the supply route less predictable.

The acquirer role can protect against supply chain threats by employing a comprehensive defense-in-breadth security strategy. Red teams can be used as a means of leveraging the acquirer's overall assurance processes. The aim of the red team is to identify potential pathways or opportunities for adversaries to exploit deficits or weaknesses in supply chain processes. The acquirer should also use operational security methods to reinforce and extend the range of protections of an active product supply chain. Operational requirements, technical requirements, and mission/business rules must include requirements for supply chain assurance. The additional requirements will help ensure that all relevant supply chain organizational elements have thought through their needs for supply chain assurance up and down the supply chain itself. These requirements should aid in reducing opportunities for unauthorized exposure or access of critical elements or processes.

The requirements themselves normally take the form of procedural recommendations that involve both technical and managerial measures. Both types of recommendations need to be maintained as a dynamic set, adjusted as needed throughout the element's life cycle. The aim of the technical measures is to ensure explicit protection of production, assembly, packaging, delivery, testing, and support activities up and down the supply chain.

Operational procedures ensure appropriate identity management, access control, and configuration management in all activities within the supply chain. As we said earlier, to ensure operational security, the identities of all the participants in system development, customers, and deliveries need to be protected through standard information assurance methods. These methods are verified through static testing processes and formal acceptance testing. One of the most effective ways of doing this uses simulations and "what if" scenarios to assess potential risk from common attacks on the supply chain process. This allows the organization to "think through" all the necessary managerial and technical measures that might be required to protect the supply chain throughout its life cycle, including security considerations in component production, assembly, packaging, delivery, testing, and support. What-ifs also allow the organization to minimize chances for unsanctioned access to critical elements or processes within the entire supply chain.

Operational procedures are also required to control product configuration and keep it in a state of constant knowledge/understanding. Secure sustainment activities include large processes such as standard configuration management as well as tailored security acceptance and deployment practices. The aim is to continuously control the supply chain in a way that the loss or compromise of product integrity is prevented.

Control frequently involves the use of audits to enforce the security of each supplier/integrator's individual procurement practices, including the way the supplier/integrator validates components "as produced," and "as received." The aim is to assess each product component, not just the product provided by the eventual supplier, to certify its trustworthiness. Therefore, every critical component of the eventual product must be vetted and the specific source of the component determined. This is not merely a case of identifying a corporate or organizational source, since most organizations encompass diverse operations whose reliability can be inconstant.

The Supplier Role

The supplier role is twofold. In simple terms, it provides a product to a customer, typically under the provisions of a contract. However, the role of the supplier becomes a lot more convoluted within a supply chain where suppliers provide product from the bottom of the product decomposition tree to the top. Normally, suppliers receive parts from subcontractors and integrate them into a larger product that is passed up the supply chain to the next level. Thus, the supplier is also a customer and an integrator. For the purposes of this discussion, we will focus strictly on the actions that are part of the basic role of the supplier: providing a product to a given customer, avoiding any specific attention to the admittedly requisite acquirer and integrator actions.

The purpose of decomposing a product into critical processes and components is to simplify the understanding of the product and break the construction process into workable steps to meet requirements as specified by the contract. The steps are the discrete set of practices by which the product is built. For the sake of clarification, the term "processes" includes all instances of product development activity for all of the components in the build. These activities are generally executed concurrently at a given level of decomposition within the development hierarchy and documented as part of the product life cycle process. As the build evolves from the bottom to the top of the hierarchy, a comprehensive and detailed inventory of the components that are integrated into the final product is maintained, as well as a listing of critical aspects of those components that might impact security. That includes a listing of

- Any identified human activities or design weaknesses that could become exploitable vulnerabilities in the final product
- The explicit testing methodologies that will be employed to verify component reliability
- Any components that will require explicit certification of correctness, such as processes handling classified material
- Any currently identified vulnerabilities and their resolution including waivers of responsibility issued by the acquirer in cases where the threat is accepted rather than mitigated
- The methodology that will be employed to certify that no counterfeit elements have been integrated into the final build

In addition to the standard assurances of process correctness, it might be necessary for suppliers up and down the supply chain to document in the specification a formal quality assurance process that includes all activities undertaken to ensure compliance with functional and nonfunctional requirements. It is important that the documentary results of this process analysis be distributed to all interested stakeholders and relevant third parties to enhance coordination and control of the process up and down the supply chain.

With respect to the actual component production, the supplier should limit complexity in product design and development. The supplier should also document and report up the supply chain to the acquirer any component vulnerabilities including those that might be discovered during routine development, patching, or upgrading. The product must always be sufficiently robust to perform reliably and enter failure mode in an appropriately controlled fashion. The supplier should undertake tests to ensure that the knowledge of potential failure modes and effects on various associated components will be passed to the acquirer.

Finally, there is the matter of ongoing product risk assessment. In general, the supplier should conduct routine assessments of all potential risks involving their areas of responsibility in the product construction process including the evaluation of all relevant components being passed up from a lower-tier supplier as well the contributions of any legacy suppliers that might be included in the product development process.

The evaluations should identify and itemize all the potential areas of risk and exposure of all components from all relevant suppliers at a lower tier in the supply chain. The supplier makes use of threat analysis techniques, misuse cases, and threat models to pinpoint potential design vulnerabilities. Then the supplier assesses the effectiveness of the protective measures that have been put in place to protect the target of assessment from any relevant threats. This analysis is carried out both by the component supplier and the organization that integrates the component at the next level in the supply chain.

All of this is necessary to ensure a uniformly correct component as it is integrated and passed up the supply chain. Additionally, new or overlooked vulnerabilities can be dealt with before they are lost in the larger product. This understanding will also help to ensure continuity of business operations. Process breakdowns can be addressed by defining a set of criteria for identifying critical elements or steps in the component construction process. Those criteria should include

- The threshold of acceptability for component performance, in terms of process performance or product acceptability
- The threshold of acceptability for authorization of rework, or modification of the original component: mean-time-between-failures (MTBF) is the measure generally used to determine this for virtual components
- Any known environmental restrictions or potentially harmful threats originating in the larger conduct of the construction process

The need to benchmark the performance of the supply chain implies the requirement that component performance and failure rates are periodically verified against benchmarks by formal tests and inspections. This requires the supplier to develop and implement a plan for routine measurement and maintenance of all components as they move up the supply chain. The aim is to protect the emerging system and its components from unauthorized alteration and security exposure through an operational set of configuration management controls. Configuration management testing and control methods apply to the technical components of the product as well as to the processes and techniques used to ensure the constant stable state of the build.

The recommended method for ensuring this is to employ defensive design in the construction of all components that are moving in the supply chain. Defensive design is the standard practice for forestalling failure or misuse by identifying all of the ways a component might be compromised. Then the component can be built or configured to minimize all known undesirable outcomes, as well as to prevent potential component failure in the long term.

Defensive design techniques are normally used by integrators and suppliers to ensure the integrity of the individual components in the supply chain. It is important to note that even when elements originate from trustworthy suppliers, these elements may still have intentional or unintentional vulnerabilities. Therefore, defensive design techniques should be utilized by the integrator to reduce risk or damage to elements and systems. To ensure this, defensive design criteria need to be incorporated into all technical requirements. The resulting requirements should provably ensure quality components.

Acquirers and integrators need to employ adequate methods to prevent unauthorized or unmonitored access to their production processes. In a supply chain, many processes and environments are highly distributed; outsourcing in a global supply chain makes such distributed approaches increasingly common. Failure to control physical or logical access to the acquirer, integrator, or supplier production environment may result in the sabotage of systems, elements, or processes; the introduction of counterfeits or malware; the theft of critical materials and information (hardcopy or electronic data); or the subversion of systems in which the elements are embedded. Adequate protection should be determined, by the acquirer and integrator, based on the threat and risk to the acquirer's mission. This is best accomplished by including the need for intellectual property security, physical, information system security, and personnel security measures in the contract. These measures must be sufficient to minimize unauthorized exposure of access to systems, elements, supply chain processes, and sensitive technical or mission/business process information of the element(s) and system(s) into which it is embedded.

The Integrator Role

As mentioned previously, the integrator has both acquirer and supplier responsibilities. In this respect, both supplier and acquirer tasks are rolled into the integrator role. The activities of the integrator function essentially comprise all levels of the supply chain. The integrator is solely the supplier in two places: where the prime contractor transfers the product to the customer and where the initial component developer starts its journey up the supply chain. Consequently, in many respects the integrator role invokes SCRM requirements in their purer sense.

Because the integrator is essentially transferring a product, it is essential that all aspects of the virtual and physical security requirements have been satisfied within the integrator's area of responsibility. This includes every facet of component development, including any necessary research, design, construction, or assembly, packaging, testing, delivery, and support; it invokes all of the practical elements of good access control. The integrator must have auditable proof that the right array of electronic, personnel, and physical access controls have been designed and implemented to provide sufficient protection commensurate with the importance of the services provided or the component procured.

Thus, as the first step in the integrator function, the integrating organization will review or audit its SCRM practices. This is often done against the stipulations of the contract with respect to the secure acquisition of the component from a lower-level supplier. Then as the component is passed up to the next level, the acquirer will review the correctness of the product with respect to security along with the prospective sustainment agreements with next level integrators. The integrator must provide a complete development/maintenance log in the form of the component's audit trail and change management baseline ledger. In cases where criticality is a factor, it might also be necessary to receive third-party certification of correctness. When lower-level COTS components have been integrated, it might be necessary to obtain this certification from the original equipment manufacturer. In that respect, it should be confirmed that the integrating organization has screened any COTS components for supply chain risks, including malware and counterfeits. If there is extensive utilization of lower-level COTS products, it might be necessary to establish a service agreement with the suppliers of those products, especially for critical components.

Obviously, when integration takes place it is necessary to adopt standard approved methods for examining and testing all lower-level components for instances of malware before performing the actual integration. This is also true for the application of changes or patches to any later updates to the product. Generically, component correctness is ensured by reviews and tests of each element integrated into the component. The aim is to find and repair any potential weaknesses and vulnerabilities. The reviews themselves can comprise standard walk-throughs and inspections, or independent audits, depending on the criticality of the product. There are resource implications in every one of these cases so it is important to factor in cost and time when planning for the actual assurance of the integrated entity.

Information and Communication Technology Product Assurance

The implementation of formal information and communication technology product assurance practices is a primary responsibility of all participants in the supply chain. The processes are the source of data for both short-term assurance during the construction phase and long-term quality/security control of the product as it is integrated up the supply chain. Testing should be both static and dynamic. Both analysis approaches are necessary to vet potential system elements prior to their integration into the final product. On the physical side of the process, once the component is complete it should be subjected to red-team and penetration testing to confirm robustness of the actual code as well as to detect any lingering defects. The level of rigor should be at the same level as final acceptance testing for the product, as specified in the contract.

In its role as the acquirer, the organization must dictate well-defined and repeatable processes for acquisition, delivery, and acceptance of the materials to be integrated into the end component. These processes are normally based on the activities and tasks of the ISO 12207:2008 Acquisition process as defined in that standard (as stated in Chapter 2). Obviously, there should be an associated, systematic vetting process to examine and accept all deliverables from lower-level suppliers. The aim is to ensure the complete correctness and freedom from counterfeits of any lower-level component received for integration. Therefore, in its capacity as acquirer, the integrator must inspect and accept all incoming items to ensure against the incorrectness, alteration, or counterfeiting of any component delivered for integration. The integrator audits and controls its product as it undergoes development with the aim of detecting and correcting all malicious activity. This also implies that all necessary documentation will be maintained throughout the shipping and receiving process. This entails the creation and systematic execution of a defined set of security audits and controls, the outcomes of which are maintained in a ledger of security information about the product. In addition to simple inspections and tests, the integrator needs to factor in any known adversarial strategies, methods, processes, and tools as well as evaluate any alternative tactics that might be employed to subvert the supply chain environment.

The primary method of attack is the inclusion of functionality that is not part of the contracted set. In general, this is malware, but it could include backdoors, logic bombs, or Trojans. This unnecessary functionality is normally designed to create unauthorized access or exposure to the system. COTS products are particularly vulnerable in this respect because they are essentially a black box to the purchaser; in many cases, COTS elements are designed to sustain a range of functionality.

It is important for the integrator role to demand that the supplier deliver components with assured security; e.g., components with trust built in. Logically, that means that suppliers should be forced to disable or remove any unnecessary functions of a delivered component, even if those features are provably harmless. Since sustainment begins when the lower-level component is accepted by the integrator, the integrator must be able to maintain, update, and patch the components to be integrated into a higher-level product. Therefore, there should be a formal sustainment process in place to manage supplied components throughout their integration. Patching will take place during varying stages of the product development process, which may provide opportunities for subversion. The sustainment processes throughout the life cycle should limit any adversary's prospects for unsanctioned access to the components or the operational process of integration.

Finally, adding requirements or unconstrained requirement changes can create new or increased supply chain risks. Each additional requirement may add elements, element features, interfaces, or new interdependencies or processes, resulting in additional suppliers or complexity. Adding additional system or element features typically increases complexity, which in turn creates opportunities for exposure and potential new vulnerabilities.

Consequently, it is important to include a provision in any contract for logical negotiation of any prospective changes in requirements. The aim is to ensure that supply chain activities will continue without the need to terminate or restart the project. This is a matter of priorities; stakeholders must be able to prioritize the incorporation of new features or a delay in adding features. It also needs to allow stakeholders to assess any proposed changes in the light of resource and timing constraints. The aim is to reduce the introduction of new vulnerabilities that might create increased opportunities for subversion.

Adopting a Proactive Approach to Risk

It is important for all stakeholders to take a proactive approach to risk up and down the supply chain with the aim of ensuring that all necessary defensive measures have been deployed. The general objective of this scrutiny is to validate compliance with all security plans and contractual obligations, ascertain that each delivered product behaves in a well-understood and predictable manner during use, and detect and classify any detected weaknesses or vulnerabilities within the product, its processes, or the electronic elements it comprises. This kind of proactive examination will allow for much greater assurance control of the product as it moves up the supply chain.

Additionally, it will help decision-makers decide whether remedial actions might be required to address any identified defects, vulnerabilities, or weaknesses up and down the supply chain. Because supply chains exist at multiple levels, assurance testing should be conducted within a range of contexts both small and large scale. Testing may be done at various phases as the product moves up the supply chain. However, the level of rigor must always be uniform. At higher levels in the process, integrators should perform thorough and rigorous analyses of all component subsystems, and processes. The aim is to identify and remediate any weaknesses that might have been passed up the supply chain from a lower level in the integrated product.

A wide range of tests might apply, including static and dynamic analysis and penetration testing. These three types of proactive approaches are dictated by a comprehensive testing plan that implements comprehensive analysis of the security and correctness of the product. Static analysis entails human review of the static architecture and code, including an audit of the evolving configuration of the system. Dynamic analysis exercises the component, using test input to determine whether component performance meets stated expectations.

Dynamic analysis can be used to evaluate elements in their environment. Dynamic analysis includes functional simulation, network vulnerability analysis, vulnerability scanning, and protocol analysis. This could include evaluating network behavior such as an unprogrammed interrupt, the unauthorized opening of network port, or erratic file system behavior such as reading or writing information to unknown or unauthorized files. A final, often ignored, feature is the need to audit to ensure that all pertinent standards and regulations are followed.

Penetration testing requires humans enabled by tools to execute or simulate attacks in one or more scenarios to detect problems. Of course, the penetration testing targets themselves must be realistic, and the simulation should mirror the adversary's known tactics, techniques, procedures, and tools. This is done by familiarizing all related personnel with known adversary tactics, techniques, procedures, and tools for attacking common security weaknesses and vulnerabilities, including keeping virus/malware signature patterns updated.

Supply chains comprise a complex, hard to understand and manage array of components that are normally interconnected in a multitude of ways to achieve a wide range of purposes with the major enterprise products and global service efforts. Moreover, large global supply chains are in a constant state of flux; as new product versions are created, technology evolves and the business environment responds to competitive pressure. Consequently, to ensure that a continuously correct state of knowledge about the configuration is maintained it is necessary to operate a well-defined security configuration management process. This is not a new thing—it is simple configuration management as practiced in the industry for the past 50 years. The aim is to ensure that all supply chain processes are evolved and enhanced to ensure against supply chain risks.

This is particularly essential during the integration phase, when different components are incorporated into a larger system. Thus, all related configuration management information must be kept in a careful and uniform fashion to ensure the continuously correct state of the evolving product throughout the integration and supply deployment process.

The integrating organization needs to develop and maintain a consistent, well-defined, and commonly understood set of configuration management policies and procedures. The means for developing and maintaining those procedures should be included in the provisions of the contract and specified in explicit program plans. Each integration project needs to be controlled by a formal Configuration Management throughout the integration system development life cycle (SDLC). Items placed under configuration management normally include the requirements set and related interface specifications, all design libraries, development tools including testing apparatuses, technical data, and information about the form of SDLC processes.

People, the Weakest Link

The performance of personnel in their various roles up and down the supply chain is critical for the maintenance of robust SCRM capability. Human behavior is both unpredictable and almost impossible to control, and it is particularly important to ensure that a well-defined and commonly understood set of personnel security controls is formulated into a baseline to control personnel-related threats and risks. These controls normally include specific efforts centered on security awareness and training, as well as implementation of the Saltzer and Schroeder (1974) *First Principles* such as separation of duties and least privilege.

First and foremost is the need to establish organizational policy and general contractual agreement with respect to awareness, education, and training at all stages of the integration process. This requires that the relevant supplier organizations define and explicitly specify access control limitations throughout the supply chain on individuals who might be able to impose adverse effects on the process. Normally, this is enforced by rigorous supply chain security awareness, education, and training programs for acquirer, supplier, and integrator personnel. This also includes designing the production process so that there is no "single personal point of failure" in the development and sustainment processes within the supply chain. This is partly a case of ensuring that the roles are properly defined, but it also requires the supplier to establish and enforce the requirement for personnel security reviews and assessments or performance among supply chain personnel. These reviews and assessments should include any individual who has exposure or access to product components, production processes, or business activities that would allow unauthorized access to the supply chain.

The integrator also needs to document that individuals who are assigned specific roles throughout the supply chain are governed by least-privilege access to project information. The integrator should be able to document that the principle of separation of duties is enforced across the supply chain. The aim is to prevent a single individual from obtaining unauthorized access to components or processes that could be compromised. This is always based on good basic identity management, access control, and process monitoring, which are the mainstay of conventional organizational cybersecurity practice. The aim is to permit timely detection and classification of anomalous behavior by any member of the organization. Therefore, it is important for all integrators to be able to document the form of their identity management, access control, and process monitoring capabilities.

The acquirer should always be able to see and judge the hiring and personnel policies and practices of integrators and potential suppliers. The aim is to be able to assess the strengths or weaknesses of the personnel security policies and procedures of all the members of the supply chain. To do that, the acquiring organization needs to be able to review and assess the personnel security policies and practices of all integrators, as well as evaluate supplier and integrator identity management and access control policies, procedures, and practices. There should be a capacity for the acquirer to assess all of the members of its supply chain for internal controls over allocation of tasks and activities and detection of anomalous behavior.

Because most supply chain failures are attributable to human error, there needs to be a means that will allow the integrating organization to evaluate pertinent personnel for competency as well as reevaluate key personnel on a periodic basis. The integrator should be able to continuously monitor its internal management decision-making with respect to the allocation of tasks and activities to a given set of pragmatic roles. Furthermore, the organization should be able to test the effectiveness of its internal controls with respect to the detection of anomalous behavior among staff as well as facilitate timely intervention to prevent or reduce adverse consequences.

Since people are the key to any supply chain risk mitigation strategy, there should be extensive and careful planning for enhancing the individual knowledge, skills and abilities (KSA) of every supply chain member through a comprehensive awareness and training program. The acquiring organization needs to be able to ensure that all of its supply chain members have developed comprehensive awareness and training programs that promote SCRM policy and procedures, as well as standard, acquisition, and procurement requirements. Integrators and suppliers should provide periodic documentation that they have implemented a comprehensive SCRM training function as well as provide periodic monitoring updates on the status of employees with respect to fulfilling contractual requirements for knowledge and capability.

This capability is built around the collection of general supply chain information as the process is executed and the accumulation of the resulting lessons learned, which should then be shared among all member organizations in the supply chain. That information should lead to targeted training to help integrator personnel identify and deal with any problems arising in supply chain work. All relevant knowledge is shared across the entire supply chain and life cycle, including new personnel, personnel who might be starting new roles, or even affected personnel from another organization.

Chapter Summary

Commercial products and services that an organization acquires are a potential source of risk. Therefore, products, whether developed or COTS, should be managed through multilayered, multivendor, and even multicultural team approaches. All the activity at every level in the process must be fully coordinated and controlled up and down the development process to ensure trust in the eventual product. Coordination of multifaceted elements of work like this requires a commonly recognized and accepted conceptual model of coherent control processes and control activities. The aim of that model is to fully encapsulate and relate the processes and activities a set of managers will utilize to understand the precise security status of any given product as it moves up a supply chain from initial design to final product integration, testing, and assurance. The ever-increasing reliance on globally sourced hardware and software components leverages the growing importance of the assurance of product trust. Globalization is an enduring fact of life in our era, and it is likely to have increasing influence over time. Thus, all organizations need a standard capability to assess and manage supply chain risks and ensure a trusted supplier base. The problem is that failures in the supplier community can cause weaknesses in the product that are beyond the control of the customer organization. Those weaknesses can be exploited, so the need to identify, assess, and specifically mitigate information and communication technology product supply chain defects is crucial to overall product trust. Mitigation is done by implementing a relevant set of electronic and management controls.

Given the requirement for stable and continuous coordination of the customer–supplier relationship, a fundamental set of commonly adopted processes needs to be planned and implemented to uphold that relationship, including the deployment of all necessary governance, business management, operational and human resources, and management controls to establish robust protection against attacks on the supply chain, whether these threats are intentional or accidental.

Most modern information technology operations are built around sourced products or services. The suppliers of those products and services provide critical components of the customer operation including software, hardware, processes, or human resources. The purchasing process must inevitably establish a defined set of associations between the acquiring organization and a given set of suppliers.

Nevertheless, the acquiring organization will most likely not be able to monitor, or at least directly control, the production and delivery processes for the products they are purchasing. Therefore, potential security risks lurk in every relationship between an acquirer and supplier base. Therefore, a requisite level of trust between the acquirer and its supplier organization should be established and assured by the definition and implementation of a formally established, commonly accepted, security governance process, one that contains a well-defined set of auditable and effective controls.

In the case of the supply chain itself, each inherent customer–supplier relationship is designed to accomplish a specific set of business goals. The number of such relationships is likely to grow with a major product or service, with the creation of a supply chain, and with a couple of customer–supplier levels down to the outer boundaries of expansion, which is conventionally set at five levels. Visibility and control are hard to maintain at any level further down the chain than the direct, e.g., subcontractor, level. This lack of control is likely to result in those relationships being badly managed or even unmanaged by the customer.

Controls are normally associated with assurance steps that are taken to mitigate meaningful risks in an upstream supplier's product or service. The controls themselves typically assure or control identified risks to supplier products, which can impact the supply chain security relationship further up the chain. Controls also enforce proper understanding of the operation of the overall execution of the process by enforcing visibility into the assurance operations of the other organization.

Transparency in the assurance of the supply chain's functioning is perhaps the most critical factor in the entire assurance process. Trust between any given set of entities depends on the ability to clearly see and understand the internal workings of the relationship on both sides. Therefore, documented and auditable assurance that the supplier has established adequate SCRM control is an essential part of ensuring a long-term trusted relationship. In most cases, the acquirer needs to evaluate the adequacy of the supplier's product or service controls based on a mutually agreed-upon set of criteria for supply chain security management. These criteria must mirror what the acquirer deems adequate to mitigate risks to an acceptable level.

This chapter focused on the means to develop a standardized range of jointly agreeable controls that can be deployed to facilitate effective risk management among all the communities of practice involved in a SCRM process. Obviously, the lack of an effective comprehensive and in-depth risk management process among the participants in product or service procurement process can very adversely impact the customer's ability to trust the security and integrity of the COTS product itself.

Several categories of risk management must be considered during the life cycle of a supply chain. First and foremost, there is the issue of overall management governance and control. It should be self-evident that a lack of formal, coordinated management or a weakness in the formally deployed control set will cause the participants in a supply chain to lose coordinated command and control over the actual nuts and bolts of the process.

Another obvious consequence of a failure to establish formal coordination and alignment in the risk management process up and down the supply chain is the potential for miscommunication and misunderstanding among various parties involved in the development process. Finally, with respect to the generic issue of difference in organizational cultures, without strict understanding of the terms and conditions of project work between partners, it is likely that the development or Acquisition process itself will be conducted dysfunctionally. Even worse, elements of the product or service responsibility will violate the customer's expectations regarding product security and integrity.

In that respect, the organizations should adopt a standard, organizationally accepted SCRM model to guide the creation of a documented and adequately understood set of coordinating processes that sufficiently itemize just those roles and responsibilities. The outcome must be suitably detailed to guide the conduct of supply chain risk mitigation activities. A common set of real-world procedures has to be documented, approved, and promulgated to the organization as a whole in order to ensure that the SCRM process is executed in a universal and uniform way. These procedures must assign and fully describe the participants in the risk assessment, risk analysis, and risk decision-making for the organization.

From a practical standpoint, the solution is characterized in the everyday best practices of the organization. Thus, the first step in the effort to embody comprehensive SCRM lies in the implementation of a well-defined set of standard and systematic best practice controls. Ultimately, information security risks lurk in all the nooks and crannies of the acquirer/supplier/integrator relationship. To assure a trusted product, the practices that will govern the three areas of supply chain practice must be fully and appropriately defined and implemented.

The requisite practices of all three are combined into a single approach based on a prospective risk management approach. A broad range of potential SCRM approaches can be applied to an information system or elements of an information system. Nonetheless, that application will vary depending on whether the actual organization is doing acquisition, supply, or integration work. This is an important distinction because the responses that an acquirer, integrator, or supplier would require requisite best practices that will diverge widely, and this divergence is exacerbated by the relative position of that organization in the supply chain. Thus, the organization's management and information security professionals need to select standard specified practices sufficient to mitigate supply chain risks in each situation.

The acquirer role is responsible for the design and documentation of a detailed, formal, end-to-end, product sourcing/Acquisition process. The design should embody the form and structure of the Agreement processes described in ISO 12207:2008: complete specifications of the preferred operational practices, acquisition strategies, and procurement activities to be employed up and down the supply chain.

The key objective is to obtain and maintain systematic knowledge of the activities of all participating organizations in the supporting tier of the development and integration process (e.g., the organizations further down the supply chain). Transparency is essential to the acquirer because it allows that organization to understand how the components of the purchased product or service are selected, created, tested, delivered, supported, and protected throughout its life cycle.

All supply chain activity must be controlled. Thus, the acquirer needs to design and implement a baseline of tangible security controls. These controls are implemented as a system by all three participants in the normal supply chain: acquirer, supplier, and integrator. The controls are put in place to mitigate supply chain risk. The implementation process is underwritten by a formal risk assessment and analysis of all management, operational, and technical elements of the product. The purpose of this assessment is to ensure that all pertinent elements, processes, requirements, and business practices up and down the supply chain operate in a harmonious way that protects against compromise.

The supplier role is twofold. In simple terms, it provides a product to a customer, typically under the provisions of a contract. However, the role of the supplier becomes a lot more convoluted within a supply chain Normally, suppliers receive parts from subcontractors and integrate them into a larger product to be passed up the supply chain to the next level. Thus, the supplier is also a customer and an integrator.

The integrator role has both acquirer and supplier responsibilities and comprises all levels of the supply chain. The integrator can only be the supplier when the prime contractor transfers the product to the customer and when the initial component developer starts its journey up the supply chain. Consequently, in many respects the integrator role invokes SCRM requirements in their purer sense.

The implementation of formal information and communication technology product assurance practices is a primary responsibility of all of the participants in the supply chain because those processes are the source of data for both short-term assurance during the construction phase and long-term quality/security control of the product as it is integrated up the supply chain. Testing should be both static and dynamic, and the level of rigor should be at the same level as final acceptance testing for the product, as specified in the contract.

It is important for all stakeholders to take a proactive approach to risk up and down the supply chain to ensure that all necessary defensive measures have been deployed. This kind of proactive examination will allow for much greater assurance control of the product as it moves up the supply chain.

The performance of personnel in their various roles up and down the supply chain is critical for the maintenance of a robust SCRM capability. Human behavior is both unpredictable and almost impossible to control. It is important that a well-defined and commonly understood set of personnel security controls be formulated to control personnel-related threats and risks.

First and foremost is the need to establish organizational policy and general contractual agreement with respect to awareness, education, and training at all stages in the integration process. Relevant supplier organizations must define and explicitly specify access control limitations throughout the supply chain on individuals who might be able to impose adverse effects on the process. Roles must be properly defined, and the supplier must establish and enforce the requirement for personnel security reviews and assessments or performance among supply chain personnel.

Since people are the key to any supply chain risk mitigation strategy, there should be extensive and careful planning for enhancing the individual KSA of every supply chain member through a comprehensive awareness and training program. The acquiring organization needs to be able to ensure that all its supply chain members have developed comprehensive awareness and training programs that promote SCRM policy and procedures, as well as standard, acquisition, and procurement requirements.

Key Concepts

- The SCRM process embodies three roles: acquirer, supplier, and integrator.
- SCRM is an essential element of doing business in a global economy.
- Supply chains rely on contracts to enforce trust.
- Supply chains are built through transparency created by effective communication.
- The creation and documentation of well-defined relationships is the basis for establishing trust among supply chain participants.
- Supply chain management should be designed, coordinated, and enforced through a single controlling entity.
- The Agreement processes as specified in ISO 12207:2008 serve as the basis for defining a formal Acquisition process.
- Trust is ensured by standard, commonly accepted controls.
- Standard control frameworks define the best-practice basis for creating SCRM.
- People are the weakest link in a supply chain.
- Staff capability should be certified and ensured by all parties in a supply chain. This is done by means of formally designed and deployed awareness, training, and education programs.
- Testing and inspections are important sources of data about the evolution of the product as well as a means of creating and ensuring continuing trust.

Key Terms

acquirer: the entity in the supply chain process who purchases a given product or service

best practice: the execution of a set of behaviors that are both well defined and commonly accepted as correct in each community of practice

community of practice: a known and commonly accepted collection of organizations all performing a given product development or service activity among which a set of formally defined relationships exists

consumer: designates the final customer in the Acquisition process, the entity for whom the product was built

control framework: a comprehensive set of standard behaviors intended to explicitly define all required processes, activities, and tasks for a given field or application

controls: technical or managerial behaviors put in place to ensure a given and predictable outcome

counterfeiting: the insertion of a component or the sale of a product that has not been produced by the manufacturer of record—can lead to unexpected failures of components and products

outsourcing: the letting of subcontract to a third-party organization to provide a product or service

physical supply chain: the action of formally delivering a specific part or module to another entity in the supply chain

process design: the act of translating product development into specific steps along a timeline for an intended, and verifiable outcome

sourcing: the organizational process of product acquisition either through a development process or by the purchase of a COTS solution

transparency: full and complete understanding of the execution and outcomes of a process among a defined set of stakeholders or partners

References

International Standards Organization (ISO), ISO/IEC 12207:2008—Systems and software engineering—Software life cycle processes, ISO, 2008.

National Institute of Standards and Technology (NIST), NIST 800-181, Supply chain risk management practices for Federal Information Systems and Organizations, 2016.

Saltzer, J. H. and Schroeder, M. D. The protection of information in computer systems. *Communications of the ACM*, 17(7), 388–402, 1974.

Chapter 4

Risk Management in the Information and Communication Technology (ICT) Product Chain

At the conclusion of this chapter, the reader will understand

- The proper methods for security categorization for supply chain systems
- The tasks that ensure adequate security control selection within the ICT product chain
- The security control implementation process
- Proper procedures for assessing implemented security controls within the supply chain
- The impact that authorization has to the effectiveness of security within the supply chain
- The generic practices of supply chain security control continuous monitoring

Introduction

By definition, *risk management* involves a systematic architecture comprising all of the necessary controls to prevent unauthorized use, loss, damage, disclosure, or modification of organizational information. In order to adequately control such an architecture, a well-defined risk management process must be in place to provide assurance that each aspect of the architecture is properly implemented and governed.

Generally, organizations design, implement, and follow a structured set of activities and tasks to establish a persistent operational risk management process. This design and management process, first and foremost, is a strategic activity in that it involves long-range considerations from an organization's operational perspective. Factor in the element of supply chain risk and the scope of the strategic activities increases exponentially. Thus, planning for strategic risk management is necessary at an organizational level, considering the implications of the supply chain, in order to ensure continuous product risk assurance. Likewise, a formal strategic planning process is necessary to implement supply chain risk management (SCRM). Risk management itself must incorporate all of the elements of the supply chain within its scope, and the process should reach the boundaries of the acquiring organization and each supplying organization.

The outcome of the implementation of a SCRM process is a concrete risk management scheme that provides a balance between long-term risk control policy, and real-world conditions and constraints. At the core of the process is a set of substantive controls that ensure the requisite level of assurance against loss. Further, those controls should be traceable directly to the policies defining their need. This is a closed-loop process in that the ongoing alignment of risk controls to policies fine-tunes the evolution of the substantive risk management process and ensures its effectiveness in all operational settings within the controllable elements of the supply chain.

Specifically, the process of risk management must include identifying and controlling information as it is created within the supply chain, risk identification (examining, documenting, and assessing the security concerns represented by a given component within the supply chain), a risk-control process (applying controls to reduce identified risks) and prioritizing their importance. It is hard to ensure against threats to the components of an evolving product because the development process is normally dispersed across a number of organizations at various levels of integration. This is potential risky because any breach of the product development chain can compromise the entire product. In Chapter 1, we used the term "weakest link." Managing the implications of the weakest link must be integrated into the activities that comprise the risk management process.

Organizations must also factor in risks associated with the use of offshore development of commercial off-the-shelf (COTS) products (over which the organization has no control). Work across organizational boundaries as defined by agreement is the basic approach to the development of most complex technology products, but most of these relationships are undefined. Software in particular is intangible and dynamically changeable. Thus, it is almost impossible to get an exact understanding of product status as it moves up the development chain.

Consequently, explicit and trustworthy risk control processes have to be applied at all levels of the supply chain.

The problem faced by organizations up and down the supply chain is that the term "risk management" is rather nebulous. To make the practice standard, the overall process itself requires a concrete statement of what a SCRM process comprises. Standardization is important because a lack of effective, coordinated implementation and execution of the elements of the process has made overall risk management efforts ineffective.

The lack of coordinated action, even within the processes of a single organization, has been so pervasive that a logical response seemed to be the formulation of a comprehensive and coherent specification of the commonly accepted best practices for risk management. That specification could then be used to guide the creation of an effective risk management scheme for all organizations throughout the supply chain. Steps were taken by the Federal Government to formally research and develop a standard and comprehensive risk management process. In this chapter, we use that standard and process as a means to demonstrate an effective risk management scheme and the ability to be extended to apply to the underlying principles of SCRM.

The specification of commonly accepted standard processes is the role of the National Institute of Standards and Technology (NIST) Risk Management Framework (RMF). Of specific interest here, NIST has developed and published a formal reference model for the management of risk: the Risk Management Framework. The framework has been written from the perspective of managing risk within an organization in isolation of risk imposed by suppliers of acquired products and services. This large-scale standard model serves as the specification of a fundamental process for understanding the risks involved in assuring information and ICT organizations, the foundation for deploying the common control mechanisms required to manage the risks within them, as well as a mechanism for applying the process steps throughout the supply chain. It has the additional advantage of providing the umbrella definition of the process for achieving Federal Information Security Management Act (FISMA) certification. The remaining sections of this chapter introduce the steps of the RMF (Categorize, Select, Implement, Assess, Authorize, and Monitor); within the context of each step's activities, we present implications of the underlying practices of successful SCRM. It is important to note that most of the steps outlined in the RMF correlate, in one way or another, to a management control that organizations must have formulated to provide adequate assurance of proper security management practices.

Supply Chain Security Control Categorization

In our discussion of the Agreement processes of ISO/IEC 12207 earlier in the book, one of the first activities required of the acquirer is the understanding and documentation of the requirements of an Information and Communication Technology (ICT) project. Many such requirements include those that protect the system and the data contained within. This understanding and documentation of security requirements involve a systematic progression through tasks aimed at identifying and prioritizing implications to underlying system confidentiality, integrity, and availability.

While it is easy to recommend that an organization proceed through security identification and prioritization activities, we must keep in mind that many ICT projects utilize services of third-party vendors and that some are local, while others must be accessed across the Internet through cloud-based infrastructures. Assessing security requirements, therefore, must address such supply chain scenarios and their associated risks. We noted in an earlier chapter that the supplier has the responsibility of justifying its capabilities. Regardless of the nature of the provided product or service, visibility of ICT security must be considered an important part of establishing a supplier relationship to ensure that the information security risks to the acquirer's information and information systems are properly managed. In order to identify and manage these information security risks, the acquirer must obtain assurance that the supplier has implemented adequate information security management and controls. When these are not negotiable, the acquirer should select a supplier's product or service based on criteria that include requirements for information security management and controls to avoid or mitigate risks to an acceptable level.

The process of understanding the acquirer and supplier business requirements while matching them to the properties of confidentiality, integrity, and availability and measuring each requirement for the degree of security risk it imposes is often referred to as "Security Impact Analysis". In an attempt to provide a checklist process to help organizations identify the level of protection a system requires, the first step of the NIST RMF, as detailed in *NIST SP 800-37 Rev 1 Guide for Applying the Risk Management Framework to Federal Information Systems: A Security Life Cycle Approach*, includes the Security Categorization process. This process addresses the need for acquiring and supplying organizations to do an initial assessment based on system information types and the organizational objectives that each supports. The RA-2 control within the RMF NIST SP 800-161 (which NIST developed specifically for the purpose of implementing SCRM) stipulates that the organization must

- Categorize information and the information system in accordance with applicable federal laws, directives, policies, regulations, standards, and guidance within its industry
- Document categorization results (including appropriate justifications) within the security plan
- Implement procedures for review and approval (by the authorization official) of the security categorization decisions

The Security Categorization step of the RMF process is considered the most important because it draws upon both organizational (acquirer and supplier) mission and goals as a means of defining ICT security activities. Throughout this step, *FIPS 199 Standards for Security Categorization of Federal Information and Information Systems* is used to define requirements for categorizing information and information systems throughout the supply chain. Furthermore, *NIST SP 800-60 Guide for Mapping Types of Information and Information Systems to Security Categories* provides the guidance the acquirer and supplying organizations need to assess the importance and sensitivity of each type of information and the information system from which it is input, processed, transmitted, stored, and shared.

The anticipated outcome of this step is that each system and information type is provided an appropriate impact level of low, moderate, or high (based on the criteria defined in FIPS 199) and is further used to select a set of baseline security controls for each information system within the supply chain, from *NIST SP 800-53 Recommended Security Controls for Federal Information Systems*, which is then customized to better meet the security needs identified for each information system. Moreover, each system's impact level determines how aggressively each organization throughout the supply chain must apply the remaining steps in the Risk Management Framework, including the assessment of security controls.

Regardless of the impact levels defined at the outset, organizational management up and down the supply chain has the responsibility to work collaboratively to ensure that security categorizations are reviewed on an ongoing basis and communicated to suppliers or other acquiring organizations in order to help ensure that the identified impact levels continue to reflect the mission and objectives of each participating organization within the acquirer–supplier relationship and its original environment. To that extent, NIST SP 800-60 suggests that security categorization routinely be revisited as an organization's mission and business functions change, because it is very likely that potential impact levels or even information types will change as well. As impact levels and/or information types change, communication mechanisms must be in place to provided suppliers and acquiring organizations the information they need to make the necessary changes to their own infrastructures.

Based on the premise that organizations enforce repetition and each organization participating in acquirer–supplier relationships actively perform system categorization, four key activities of the process must be performed by each participant as shown in Figure 4.1:

1. The acquiring organization as well as each supplier must develop policies pertaining to information system identification for the purpose of security categorization. The system generally includes a security boundary, which is often used in ICT discussions to describe the system or part of a system that is under a specific organization's administrative control. It is worth noting that, while an acquiring organization has little administrative control over supplier administrative policies, the acquirer *does* have control over which

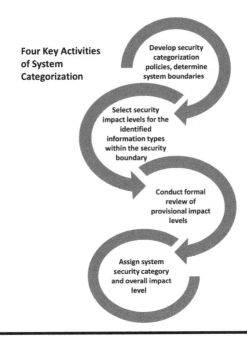

Four Key Activities of System Categorization

Develop security categorization policies, determine system boundaries

Select security impact levels for the identified information types within the security boundary

Conduct formal review of provisional impact levels

Assign system security category and overall impact level

Figure 4.1   Four key activities of system categorization.

supplier it chooses, based on what is known of the policies defined to support risk management. Therefore, it is in the best interest of all parties of the supply chain to develop such policies and make them available upon request. The deliverable for this activity is a document that clearly states each organization's business and mission areas and the identification of the information types that are input, stored, processed, outputted, and shared from each system. Additionally, this document should include the basis for the information type selection. In addition to such policies being made available to acquiring organizations, they must be available to suppliers upstream from origin to completion in the supply chain. The documentation produced in this and the next three activities becomes an addendum to the organization's security plan, which becomes authorized later in the risk management process.

2. The acquiring organization as well as each supplier should select the security impact levels for the identified information types within the security boundary. Those security impact levels can be selected from the recommended provisional impact levels for each identified information type using the guide, NIST SP 800-60, Volume II Appendices C and D or from the FIPS 199 criteria for specifying the potential impact level based on security objective. The deliverable for this activity is a document that states the provisional impact level of confidentiality, integrity, and availability associated with each of the system's information types.

3. Next, each organization must conduct a formal review of the provisional impact levels for the identified information impact levels for the information types. This activity also includes the adjustment of the impact levels as necessary based on the following considerations:
   - Confidentiality, integrity, and availability factors
   - Situational and operational drivers (timing, life cycle, etc.)
   - Legal or statutory reasons

   The deliverable for this activity includes a document of all adjustments as well as the final impact level assigned to each information type and the rationale or justification for the adjustments.

4. Finally, the organization must assign a system security category and overall impact level. A review is conducted to analyze the identified security categorizations for the aggregate of information types and determine the system security categorization by identifying the highest security impact level for each of the security objectives (confidentiality, integrity, and availability). The deliverable for this step is a document containing the assignment of the overall information system impact level, based on the highest impact level for the system security objectives (confidentiality, integrity, and availability).

The "carry forward" dynamic of documentation throughout the categorization process is important because the conclusions made throughout this process are used as input to the selection of the set of security controls necessary for each system and the system risk assessment. The minimum security controls recommended for each system security category can be found in several helpful resources such as *NIST SP 800-53 Revision 4, Security and Privacy Controls;* the Council on Cybersecurity, *Critical Security Controls for Effective Cyber Defense, Version 5.0*; SANS Institute, *Top 20 Critical Security Controls*; and the *Payment Card Industry Security Standards Council, Data Security Standard (PCI-DSS).*

Categorization Success through Collaboration

The underlying success in implementing the SCRM based on existing frameworks such as NIST's RMF is largely dependent on a collaborative effort among all internal and external organizational entities that are directly impacted by the way information security practices are performed. Appropriate collaboration within the supply chain is so important that, if you will recall from discussions in earlier chapters, it is directly built into many of the activities and tasks of the Agreement process of ISO/IEC 12207. Collaboration is indicative within the activities of NIST IR7622 *Notional Supply Chain Risk Management Practices for Federal Information Systems.* You will see the same is true of the activities defined in NIST SP 800-161. Through a collaborative effort, senior management is able to be proactive in making security risk decisions that have an impact on each organization's ability to achieve its mission, and vital business operations remain functional while maintaining an adequate level of security within each of those operations. Security teams throughout the supply chain must continue to communicate with each internal and external business entity dependent on information and the systems that provide the capacity for its processing, storage, and transmission in an effort to provide the amount of guidance and direction necessary to achieve success within the categorization process. Such outreach activities include coordinating the definition and distribution of organization-level information types, leading organization-wide and supply chain categorization sessions, providing training to ensure the categorization process is completed according to directives, and developing templates or obtaining tools to provide assistance in the completion of categorization activities.

Additionally, the security teams must continue their effort in developing and maintaining relationships associated with enterprise architecture teams and affected system operations personnel to ensure that security policies based on the system's impact level are implemented properly, common security controls are implemented, and configuration management includes security considerations within the operational decision-making process.

> The success of the categorization process is dependent upon the collaboration among the organization's many entities. Senior leaders must balance the benefits gained from using information systems with the risks that the same systems will be the vehicle through which adversaries cause mission or business failure. Working together, senior leaders can make informed decisions, provide adequate security, mitigate risk, and help ensure the organization's missions and business activities remain functional.
>
> *(NIST, 2009)*

Supply Chain Security Control Selection

A security program, whether at the organization or the system level, should include an appropriate mix of security controls: management, operational, and technical. Through security control selection, organizations throughout the supply chain are motivated to identify the security controls necessary to satisfy each individual ICT system's security requirements as well as those throughout the supply chain. This step contains tasks associated with documenting those controls in the system security plan.

From an input → processing → output perspective, the results of the system security categorization serve as input to the selection of security controls, considering the impact level assigned to the information system (having used FIPS199 as a basis for drawing those conclusions) corresponds to a baseline set of security controls that, in combination, provide the minimum security requirements (initial baselines), defined by FIPS200 *Minimum Security Requirements for Federal Information and Information Systems*, necessary for protecting systems at each impact level. Through the selection process, each supply chain organization should use security requirements and risk assessment documentation developed for each of the components that make up the ICT system in combination with the system security categorization to identify the appropriate initial security control baseline and modify that baseline to address the needs of the system. The outputs of the security control selection process are a tailored security control baseline, continuous monitoring strategy, and an approved initial version of the system security plan that can be distributed during the RFP and bidding activities of the Acquisition and Procurement process.

Security control selection identifies all of the controls relevant to each ICT system, regardless of which functional unit or supply chain organization is responsible for providing them. Most ICT systems include a mix of system-specific, common, and hybrid security controls. "System-specific controls are security controls that provide a security capability for a particular information system only and are the primary responsibility of information system owners and their respective authorizing officials. Common controls are security controls that can support multiple information systems efficiently and effectively as a common capability. When these controls are used to support a specific information system, they are referenced by that specific system as inherited controls. Hybrid controls are security controls where one part of the control is deemed to be common and another part of the control is deemed to be system-specific" (NIST, 2011).

Based on the descriptions of these control types and within the context of the discussions of this book, system-specific controls are certainly important considerations for each individual organization. However, the selection of common controls and hybrid controls and their associated providers becomes a significant implication to the streamlining of control selection throughout the supply chain. Security control baselines defined in system security plans indicate the type for each control and, in the case of common or hybrid controls, may incorporate control information in other system security plans. Thus, all relevant documentation must be shared to establish the level of trust necessary for effective SCRM. The security control baseline defined during this step serves as the basis for security control implementation and assessment activities conducted in the next two steps, discussed in later sections of this chapter.

The security controls selected to support an ICT system typically include both system-specific controls (provided by the system or the operational and management functions dedicated to the system) and common controls provided by other systems or parts of the organization (or external organizations) that protect multiple systems. To achieve streamlined SCRM, each supply chain organization needs (prior to selecting security controls) to identify common control providers and the security controls available for their ICT systems to use and understand common controls enough to determine if they meet their own system's security requirements and, when plausible, those of known supply chain organizations. When available, common controls do not normally satisfy all of an organization's ICT system security requirements in combination with suppliers and other organizations acquiring the organization's products or services. Therefore, determination must be made whether to implement a system-specific alternative or if the common control can be partially utilized as a hybrid control.

The task of identifying common controls should be performed at the organizational level with considerations made for the security requirements identified by supply chain organizations, with a directory or inventory of controls made available to the management of all participating organizations overseeing the identification process. The ability to use preidentified sources of common controls greatly simplifies the control identification process for security team members of each organization and the management infrastructures overseeing the control selection process, thus eliminating the need to search for common control providers as part of the task and allowing attention to be focused on assessing the suitability of available controls as mechanisms for satisfying minimum security requirements. Security team members and management performing common control identification should be aware of the potential that more than one provider exists for the same control, as is often the case when more than one operating environment is available for information system deployment, thus adding the additional activity of evaluating the provider based on characteristics such as credibility, reliability, and their own security posture. Perhaps the most significant advice that can be given is that when possible the same common control provider should be utilized throughout the entire supply chain, thus providing a level of consistency and adequate security assurance.

While it is not a requirement for private sector supply chain organizations to follow federal standards and guidelines, organizations doing so begin security control selection by identifying the baseline security controls corresponding to the impact level assigned to the ICT system during security categorization. One such guideline is NIST Special Publication 800-53 *Security and Privacy Controls for Federal Information Systems and Organizations*. The minimum security requirements defined by FIPS 200 cover 17 security-related areas with regard to protecting the confidentiality, integrity, and availability of systems and the information processed, stored, and transmitted by those systems. NIST SP 800-53 provides a complete set of security controls categorized into family names that match the minimum security requirement categories identified in FIPS 200 (with the addition of an eighteenth category for project management, three security control baselines—low, moderate, and high impact) and guidelines that organizations can use to tailor standard baselines to their own specific needs according to the organization's missions and the environments in which the operations or ICT systems are performed. The established initial baselines represent a starting point for the selection of security controls, serving as the basis for the reduction or supplementation of security controls in ICT systems. Alternatively, NIST SP 800-161 provides "supply chain specific" guidelines that organizations should utilize to identify initial baselines and formulate an underlying SCRM process. While NIST SP 800-53 defines many of ICT supply chain-related baselines and control families in Appendices D and F, NIST SP 800-161 augments NIST SP 800-53 by repeating ICT SCRM-related controls, includes additional supplemental guidance, and provides new controls within each family as necessary. The ICT SCRM controls of NIST SP 800-161 are organized into the same 18 families as NIST SP 800-53, with the addition of a nineteenth ICT SCRM-specific family, "Provenance". Important to note is that NIST SP 800-161 should not be used as a replacement for NIST SP 800-53. Rather, the two should be used interchangeably to adequately select the minimum security requirements at the system, organization, and supply chain levels.

Both NIST control catalog publications provide compensating and supplemental controls for many of the security controls listed within the publications' families. In some instances, an organization may find that a baseline security control applies for a system, but implementing the control specified in the baseline is beyond the organization's resource capacity from a triple-constraint (scope, time, and cost) perspective, or its selection may conflict with supply chain organizations. Prior to deciding to accept, avoid, or otherwise respond to the threats and vulnerabilities affecting the organization by failing to implement a required control, management should consider the selection of compensating controls as an alternative that satisfies the same security objectives. These controls are designed to satisfy the requirement of a security measure that is determined to be too difficult or impractical to implement. As an organization introduces compensating controls as an alternative to an otherwise less-feasible baseline, those controls must be documented and rationale explained for choosing the alternative instead of the baseline in much the same way as is done to document selection of common or hybrid controls.

Likewise, considering system-specific controls may lead organizations to select supplemental security controls beyond the minimum requirements specified in the appropriate baseline for the system. The guidelines provided by NIST SP 800-53 and NIST SP 800-161 provide vital information for the implementation of supplemental controls and control enhancements, which organizations may choose from the requirements in a higher-level baseline or from several optional controls and enhancements in the security control catalog that are not assigned a baseline. Each individual organization must determine the necessity for supplemental controls by comparing the security requirements defined for each ICT system throughout the supply chain with current capabilities and the expected effect of implementing baseline controls. Additionally, any requirements that have not been satisfied by baseline controls may indicate a need for supplemental control considerations. As is the case for compensating controls, all decisions regarding the addition of supplemental controls or enhancements should be documented to provide supporting feasibility analyses so that management and other organizations within the supply chain can understand the basis for the control implementation.

The Eight Tasks of Control Selection

NIST has defined an eight-step process for selection of security controls. While the process was initially documented in NIST SP 800-37, the same activities should be performed when extending risk management considerations throughout the supply chain. The main difference, of course, is the utilization of NIST SP 800-161 as a guideline to enhance the controls available in NIST SP 800-53 for the purpose of supply chain implications. The eight steps defined by NIST are summarized in Figure 4.2 and explained in detail throughout the remainder of this section.

Documentation Prior to Selection

To provide the most effective approach to selecting security controls traversing across a supply chain, each organization must go through a preliminary activity involving the collection of relevant documentation specific to the system. Detailed attention should be paid to the documentation that describes the means by which ICT systems are interconnected with other suppliers or systems of other parties acquiring the organization's ICT-related products or services. In particular, the security plan, procedural documents, and assessment risk results must be made available. It may seem strange that assessments would be performed before the selection of controls. Assessments include testing, evaluations, reviews, and analyses. Each organization within the supply chain conducts assessments of ICT systems, components, products, tools, and services.

The intention is to conduct assessments in order to uncover unintentional and intentional vulnerabilities. SA-12 of the System and Service family of controls defined in NIST SP 800-161 stipulates that the organization can adopt any appropriate assessment method or combination of methods prior to selecting supply chain components used in the organization's information system or ICT supply chain infrastructure. Further, the guideline states that selection of assessment method depends on the identified impact level (depth and breadth required for component selection) and should take into consideration the underlying security requirements and budgetary constraints when making methodology decisions.

Select Initial Security Control Baselines and Minimum Assurance Requirements

At the conclusion of security categorization, each supply chain organization will have determined and documented into its security plan the appropriate impact level for the system. With that done, the organization can begin identifying the initial set of security controls and minimum assurance requirements. More specifically, a process is initiated by which the initial set of security controls is selected through the identification of the baselines listed low, moderate, and high in NIST SP 800-53, Appendix D. NIST SP 800-161, Appendix A, provides a similar selection

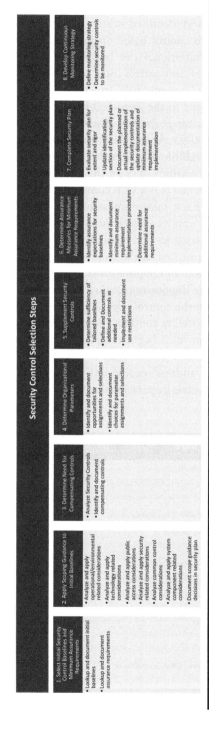

Figure 4.2 Security control selection steps.

mechanism by presenting a table containing a subset of NIST SP 800-53 controls directly related to SCRM and identified as a high baseline. Further, the table provides categorization of each control based on its established SCRM baseline and the risk management organizational tier initially defined in NIST SP 800-39. Figure 4.3 shows a sample of the controls listed in NIST SP 800-161, Appendix A.

Recall the discussion of "strategic risk" in Chapter 3. The baseline controls are carefully chosen as a requisite to the identified impact levels (based on FIPS 199) determined during security categorization. The minimum assurance requirements are defined in NIST SP 800-53, Appendix E. In general, security assurance is the means by which trust is established within the ICT system. Given the scope and complexity of SCRM relationships, it is an understatement that identification of minimum assurance requirements is a vital component to the overall supply chain control selection process. Each minimum assurance requirement is grouped by system impact level, and it applies to each control within the final set of security controls. Once the initial baselines and minimum security requirements are identified, they should be documented in the security plan. The documentation must include the identified controls and requirements in addition to the justification for having made each decision. The documentation written here will be updated after subsequent activities within the control selection process are complete.

Once an initial set of baseline controls and minimum assurance requirements is selected, the organization must begin the process of fine tuning or "tailoring" the baselines using the guidelines provided in NIST SP 800-53. Through this activity, organizations can address specific business processes and organizational requirements, constantly evolving operational environments by adjusting the initial

Control No.	Control Name	80053 Rev. 4 High Baseline	SCRM Baseline	Tiers 1	Tiers 2	Tiers 3
AC-1	Access Control Policy and Procedures	X	X	X	X	X
AC-2	Account Management	X	X		X	X
AC-3	Access Enforcement	X	X		X	X
AC-3 (8)	Access Enforcement | Revocation of Access Authorizations		X		X	X
AC-3 (9)	Access Enforcement | Controlled Release		X		X	X
AC-4	Information Flow Enforcement	X	X		X	X
AC-4 (6)	Information Flow Enforcement | Metadata		X		X	X
AC-4 (17)	Information Flow Enforcement | Domain Authentication				X	X
AC-4 (19)	Information Flow Enforcement | Validation of Metadata				X	X
AC-4 (21)	Information Flow Enforcement | Physical / Logical Separation of Information Flows					X
AC-5	Separation of Duties	X	X		X	X
(AC-6)	(Least Privilege)	(X)	(N/A)			
AC-6 (6)	Least Privilege | Privileged Access by Non-Organizational Users		X		X	X

Figure 4.3   Sample ICT SCRM control summary.

security control baselines. Tailoring activities include applying scope guidance to the initial baseline. Scope guidance involves:

■ Determining the extent to which a security control applicable to a specific information technology is necessary to a specific ICT system
■ Developing the specification of compensating security controls, if it becomes necessary to replace recommended security controls
■ Developing the specification of organization-defined parameters values, when required to implement specific security controls

The activity of applying scoping guidance entails the review of the ICT system in order to determine whether the use of common controls, physical infrastructure-related considerations, or technology-related considerations is needed. This assessment is made for each baseline security control in the ICT system. NIST recommends that organizations take into consideration operating environments, technology, physical infrastructure, public access, policies & regulations, security objectives, common controls, system component allocations, and scalability when making scoping guidance decisions.

Further, any organization within the supply chain should consider that suppliers or acquirers might not be able to offer customization or tailoring to the extent that organization determines of their SCRM needs in addition to the budgetary constraints imposed on such tailoring. With that in mind, organizations acquiring products and services of ICT suppliers should analyze the costs versus the benefits of those products and services as they make their final acquisition decisions. The acquiring organization must also keep in mind the lack of control it has over supplier decisions. As such, the supplier may choose to keep its processes or products as is and not support the acquirer's SCRM requirements. Collaborative conversations between the two parties may result in an understanding and identify acceptable solutions when such challenges occur.

Determine Need for Compensating Controls

As the organization concludes the tailoring process of the control selection, there are times in which specific controls may not be appropriate or feasible for implementation. In those cases, the organization should evaluate the appropriateness of compensating controls. Such controls are an alternative to specific controls in the low, moderate, or high baselines. These controls are intended to provide equivalent or comparable protection for organizational ICT systems throughout the supply chain and the information processed, stored, or transmitted by those systems.

As organizations make control selection decisions, they must consider that in many cases, they use external ICT service providers to manage their mission and business functions. Recall from previous discussions that the strategy of outsourcing ICT systems and services creates a set of ICT supply chain concerns that reduces the acquirer's visibility into, and management of, the outsourced functions. With just that point being considered, an organization can find itself faced with the need to enforce increased rigor in defining ICT SCRM requirements. Regardless of who performs the services, the acquirer has the responsibility for the risk imposed to its information infrastructure and data that may result from using these services. As such, an organization must implement a set of compensating ICT SCRM controls to address this risk and be willing to accept that risk. Once implemented, those compensating controls must be adequately communicated, verified, and monitored through such mechanisms as contracts, interagency agreements, lines of business arrangements, licensing agreements, and/or supply chain transactions.

Determine Organizational Parameters

Many of the security controls listed in Appendix F of NIST SP 800-53 and Appendix B of NIST SP 800-161 include control enhancements that contain system-specific or organization-defined parameters that add a considerable amount of flexibility when defining selected portions of the controls to effectively meet specific organizational security requirements and objectives. Each parameter contains a predetermined set of values that can be assigned by the organization. Once the organization has completed an initial pass through scoping considerations, and has made a selection of compensating controls, it must begin a review of security controls and control enhancements to identify appropriate assignment/selection statements in order to determine the most effective organization-defined values for the identified parameters. Once the organization has defined the parameter values and control enhancements, it becomes part of the control and enhancement. Conversely, some organizations choose the parameter values before selecting compensating controls since that activity completes the control definitions and may adversely affect compensating control requirements. After the system-specific and organization-defined security control parameters are defined, they must be documented in the security plan. When appropriate, the selection of control parameters that directly impact supply chain risk should be communicated through established mechanisms in order to promote the constancy and visibility of mitigation practices.

Supplement Security Controls

Many of the control decisions made relating to initial baseline selections and tailoring activities result from considering assessment findings done prior to the initiation of the selection step of the RMF. Those assessments range from the potential of risk at the business/mission and organization levels and extend to those risks that may be imposed through acquiring ICT products and services from external suppliers. Most organizations find that additional security controls or control enhancements are necessary to mitigate specific threats to and vulnerabilities in an ICT system or to satisfy the specific organizational and supply chain security requirements prescribed through acquirer/supplier contracts, laws, organizational policies, standards, or industry regulations.

To thoroughly understand what supplemental controls are necessary, the organization must analyze the tailored security control baseline to determine whether the controls already selected are in accordance with the organizational and supply chain security requirements. Based on an understanding of the risk assessment results; mission/business requirements; supplier/acquirer requirements; system description; and applicable laws, policies, standards, or regulations, including organization-specific guidelines, reviews can be conducted to identify potential threats, vulnerabilities, and resulting system risks as a means for better understanding the need for additional security controls or control enhancements to adequately protect the ICT system.

When it has been determined that additional security controls are required, NIST recommends that they be selected from Appendix F in the NIST SP 800-53 security control catalog. Likewise, when considering the implications of supply chain security, Appendix B of NIST SP 800-161 provides supplemental guidance and related controls (similar to what is provided in NIST SP 800-53) organizations can consider for implementation. Nevertheless, the organization must be cautious not to implement information technology beyond its ability to adequately provide protection, thereby preventing the ability to implement sufficient security controls within an ICT system to adequately reduce or mitigate risk within the organization. Additional supply chain implications become a factor when technology beyond the organization's scope of protection is warranted and an alternative strategy is needed to provide the necessary protection. Such a strategy must consider all of the additional risks imposed by the additional use of technology, both COTS and that which is operated and maintained by third-party vendors.

In other cases, the organization may adopt an alternate approach. Instead of adding additional security controls, the implementation of a security control is modified. Such modification could include increasing the frequency of security activities, increasing the level of detail provided in security documentation, increasing the scope of operating procedures, or increasing the frequency of security reporting during continuous monitoring activities.

Once the supplemental security controls have been identified, the control decision and justification for making that decision must be included in the security plan. The justification for supplementing a control should include the reasons supplementation is necessary, reference to the control catalog from which the control was chosen, and details that support the supplemental control's ability to satisfy system security requirements. With the inclusion of supplement controls, the plan should now have the final selected set of security controls.

Determine Assurance Measures for Minimum Assurance Requirements

The NIST SP 800-53 guideline devotes an entire section to the appropriately defining security assurance and trustworthiness from the perspective of security control specification, design, development, implementation, and maintenance. That trustworthiness in security requirements is reinforced from the supply chain perspective in the SA-13 control in NIST SP 800-161, which states that:

> These ICT SCRM requirements must include a clear definition of supply chain disruptions, human errors, purposeful attacks, and other risks. Processes and procedures must be defined as part of the requirements activities to ensure that not only components of the ICT supply chain infrastructure and the information system are predictably behaving, but that the processes and procedures also support the requirements for trustworthiness.

> *(Boyens et al., 2015)*

In determining assurance measures that satisfy minimum assurance requirements, each supply chain organization must establish a *measure of confidence* that security controls can be validated to ensure that they have been implemented correctly, perform the intended functionality, and meet the outcomes as specified in the security requirements. Trustworthiness provides the *measure of confidence* that supports the systems capability of preserving the confidentiality, integrity, and availability of the information that is being processed, stored, or transmitted by the systems that operate within an identified area of threat. Realistically, that measure of confidence is very difficult when an organization is dealing with a multitude of ever-changing supplier–acquirer relationships.

The NIST minimum assurance recommendations are defined in SP 800-53, Appendix E. The guideline stipulates that, for security controls in low-impact ICT systems, organizations should focus on ensuring that there are no obvious errors and that as flaws are discovered they are addressed in a timely manner. The security controls in ICT systems categorized as moderate impact should focus on actions that foster increased confidence in the correct implementation and operation of each control. It is still likely that flaws will be uncovered; however, during implementation, specific capabilities and documentation can be integrated into the control to increase confidence and help alleviate the flaws, while meeting the required function or purpose. The documentation integrated into the control becomes important when assessors must analyze and test the functionality of the control as part of the overall control assessment step of the RMF.

Security controls in ICT systems categorized as high impact should focus on expansion to *require* integration within the control the capabilities necessary to provide consistent operation of the control and continuous improvement in the control's effectiveness. Recall that NIST SP 800-161 echoes the high-impact controls listed in NIST SP 800-53 that directly relate to supply chain security. The improved effectiveness can be sought by integrating many of the related controls listed in that guideline. Likewise, during each phase of the system life cycle, it is expected that the organization will prioritize the requirement of associated design and implementation documentation to support these activities. The documentation integrated with the control is also valuable to the assessors who must analyze and test the internal components of the control as part of the assessment process.

As each control is designed and implemented, it should be reviewed to determine whether any additional enhancements or documentation is needed to satisfy assessment criteria. Appendix E of SP 800-53 provides assurance requirements for each of the system impact levels (high, moderate, and low). Those requirements define what degree of assurance an organization is expected to implement in order to satisfy assessment criteria. A table of assurance-related controls is provided for each level to identify those controls that must be implemented in order to satisfy the assessment expectations for that level. The appendix also provides additional assurance requirements available to developers/implementers of security controls that supplement the minimum assurance requirements for low-, moderate-, and high-impact ICT systems.

Complete Security Plan

As has been emphasized in this and other chapters, communication is a key component to any legitimate approach to quality SCRM. At the core of communication is documentation to be distributed to suppliers and acquires to facilitate practices of managing cybersecurity risk; the most vital of these documents is the security plan. Within the security plan are the justifications for all of the decisions made during the initial security control selection, tailoring, and supplementation processes. Organizations should provide convincing rationale for those decisions, that they directly support prescribed security objectives and requirements of the organization and those known by management, suppliers, and acquirers. The clarity of the information contained within the security plan is essential when examining the overall security posture of the ICT systems across the entire supply chain, taking into consideration the security impact on each organization's mission and business objectives. The selected security controls and supporting rationale for control selection decisions are documented in the plan.

The challenge faced by organizations is preparing a security plan that puts the organization's security objectives into a clear context and is a valuable source of reference for suppliers and acquirers. It would appear beneficial, in this case, for organizations to adopt commonly acceptable approaches to preparing the plan. NIST SP 800-18, *Guide for Developing Security Plans for Federal Information Systems* provides the guidance organizations need to develop security plans that adequately define underlying security strategies.

The security plan is scalable with regard to the extent and rigor of the implementation. The scalability is largely dictated by the security categorization of the system. The security plan for a high-impact ICT system may be very detail oriented and contain a significant amount of implementation criteria. Likewise, the security plan for a low-impact information system may be much briefer and contain considerably less implementation detail. Regardless, there are many NIST-compliant security plan templates available that organizations can utilize as a basis for formulating their own security planning strategies.

When documenting how the assurance requirements are implemented in the ICT system, the extent of the detail is usually scaled to the system's impact level, since low-impact systems require much less explanation than their high-impact counterparts. However, the plan should provide enough detail for a well-defined implementation of the minimum assurance requirements.

When providing the documentation related to security controls in the plan, organizations must be careful to include criteria related to the reductions or additions made to the security control baselines. Additionally, criteria should include detail pertaining to whether the addition or reduction is a result of security implications transpiring from within the organization or the result of actions taken by a supply chain organization. This information not only satisfies standardized definitions of the contents of security control documentation in the system security plan, but it also provides guidance to SCRM oversight and the security team responsible for implementing and configuring the security controls to satisfy the system's defined security requirements. In most instances, management, operational, and technical controls include parameters associated with policy, acceptable use, time periods, frequency of execution, or other attributes that vary among ICT systems. Selection of controls is not complete until values for these parameters have been determined and documented within the security plan at the level of abstraction necessary to support effective and efficient implementation and configuration of each control.

Develop a Continuous Monitoring Strategy

Later in this chapter, we will introduce the Authorization step of the NIST RMF and its implications to the underlying principles of SCRM. Upon successful authorization, the process of continuous monitoring of implemented security controls begins. However, as part of security control selection, a strategy is developed that adequately describes how continuous monitoring of the controls will proceed. To facilitate a clear and directed approach to continuous monitoring, we recommend that supply chain organizations adopt NIST SP 800-137, *Information Security Continuous Monitoring for Federal Information Systems and Organization,* to ensure a consistent formalized approach to preparing a continuous monitoring strategy.

That guideline states that "The goal is to provide: (1) operational visibility; (2) managed change control; (3) and attendance to incident response duties" (Dempsey et al., 2011).

As defined by NIST, the process for continuous monitoring includes, first, the definition of a continuous monitoring strategy based on risk tolerance that maintains clear visibility into assets and awareness of vulnerabilities and utilizes up-to-date threat information. Next is the establishment of measures, metrics, and status monitoring and control assessments frequencies that make known organizational security status and detect changes to information system infrastructure and environments of operation and status of security control effectiveness in a manner that supports continued operation within acceptable risk tolerances. The continuous monitoring program should then be implemented to collect the data required for the defined measures and report on findings and automate collection, analysis, and reporting of data where possible. Once an adequate analysis has been performed, it is easier for an organization to respond to assessment findings by making decisions to mitigate technical, management, and operational vulnerabilities, accept the risk, or transfer it to another authority. Finally, a review and update process should be performed in order to revise the continuous monitoring strategy and maturing measurement capabilities to increase visibility into assets and awareness of the vulnerabilities of the organization, provide continued consistency with suppliers and acquirers, further enhance data-driven control of the security of an information infrastructure across the entire supply chain, and increase organizational flexibility.

Once the continuous monitoring strategy is developed, approval is normally obtained in combination with the approval of the security plan. Many organizations take advantage of automated tools and supporting databases to conduct the continuous monitoring activities. Such tools facilitate near real-time risk management for the ICT system and provide a streamlined approach to the way security authorization activities are performed.

Supply Chain Security Control Implementation

The controls activities defined within the RMF involve applying the decisions that have been made for mitigating risk to the ICT system and its supporting processes. Each organization within the supply chain has four choices: accept the risk, transfer the risk, limit the risk, or avoid the risk, with careful consideration made to the effect that those choices have on the interactions the organization or the organization's ICT system has with suppliers and acquirers. Coming out of the Selection step, each information asset now has an assigned risk level and a chosen set of controls for mitigating risk. The most common implementation decision is to limit the risk by putting controls in place to protect an organization's information assets and systems. As the activities of implementation are performed, the organization should enforce continuous monitoring and regular updating as a way of keeping the identified risk at an acceptable level.

In essence, the tasks associated with security control implementation provide the means by which the organization is able to incorporate the controls identified and approved as part of the security plan within the functional and technical requirements identified for the system and its overall design. Organizations can select from three categories of controls: managerial, technical, and operational. An ill-advised approach to implementation would be for an organization to focus its efforts on implementing technical and operational controls while implementing the managerial controls to provide the necessary support. Organizations should start by implementing a well-defined managerial control structure based on the underlying principles of SCRM. Then it should implement the technical and operational controls utilizing that management structure as a means for ensuring that technical and operational risks identified throughout the entire supply chain can be effectively mitigated. The RMF identifies two tasks associated with implementation: security control implementation and security control documentation. In performing those tasks, organizations must be careful to consider all three control categories.

Implement the Security Controls Specified in the Security Plan

As we consider the implication of implementing security controls, it is easy to draw wrong conclusions of the extent to which this step depends upon supply chain relationships. Consider that if we were still performing ICT processes characteristic of the 1980s, 1990s, and prior, most hardware, software, and management implementation was done in house by development teams that specialized in individual areas of the ICT system or management. As time evolved, those development teams gradually included "project specific" contractors that were hired specifically to complete large-scale ICT projects. Today, most large organizations depend upon the supplier relationships they have with full-service life cycle organizations in order to outsource development. With that in mind, it becomes clear that security control implementation has significant implications for the discussions we had in an earlier chapter about the ISO/IEC 12207 Agreement process activities of the System Development Life Cycle (SDLC). That point alone speaks volumes to the importance of the existence of a well-defined life cycle process that integrates the steps of SCRM.

While each organization is unique in terms of the extent by which supply chain plays a role in security control implementation, the responsibility for completing the activities of this task is generally assigned to all pertinent areas of IT department, other affected ICT system owners within the supply chain, common control provider(s), the Chief Information Systems Security Officer (CISSO), and the information systems security engineer. The underlying objective is to implement the system's required security controls. The means by which controls with supply chain implications are implemented is largely based on the guidance provided by NIST SP 800-53A, *Guide for Assessing the Security Controls in Federal Information Systems and Organizations,* with care taken to prioritize the implementation of the selected Configuration Management, Security Assessment and Authorization, and other management control families of NIST SP 800-161.

The organization should use NIST SP 800-53A as a means for providing the specific requirements used to assess the security controls implemented in the ICT system. Further, the organization must ensure that controls with supply chain implications adequately meet the assessment criteria defined by the supplying or acquiring organizations. In those cases, in addition to the techniques described in NIST SP 800-53A, an organization may alternatively use other techniques such as supplier self-assessment, acquirer review, or third-party assessments for measurement and ensuring that controls that meet the organization's requirements are properly in place. When an organization defines ICT SCRM implementation requirements, it may discover that third-party assessments do not address all specific requirements required for proper implementation validation and verification. Additional evidence may be needed to justify implementation accuracy and specification consistency. By approaching implementation through techniques that first consider assessment, the organization can be assured that the required security controls are implemented to the same standard required when the system is assessed. This "back into" practice of implementation also ensures that the system's security controls are developed correctly and are validated as compliant during security control assessment.

One approach to ensuring the correlation between implementation and assessment is to first consider the overall structure of the organization's ICT security posture and the dependencies that exist between that organization and the supply chain partners (COTS providers or other third parties providing hardware, software, and life cycle services). Based on the architecture that security controls are built into and the availability of approved and contractually agreed-upon common controls, the organization, suppliers, acquirers, and control provider can develop methods of ensuring that the security controls within their own domains of responsibility are implemented correctly, provide the required security mechanisms, and support the ICT architecture across the supply chain and ICT security strategies. This multiorganization architecture will provide a mechanism from which specific security controls and common controls can be allocated to the ICT system, its functional units, and common control providers. The availability of a defined backup procedure, configuration management, appropriate network firewall configurations, and other security mechanisms and services are just a few ways in which a security control may be integrated into the multiorganization ICT security architecture. Such services or products are then selected, developed, and allocated, after approval and contractual agreement, to specific systems and components requiring that capability.

Consider the circumstance of a third-party provider requiring access to a segment of an organization's ICT system via firewall implementation. It is not uncommon for supply chain relationships to require specific configuration settings at the point of the firewall, with a defined upgrade schedule, on all network and telecommunication system components that are capable of having that form of network security applied to them. Such a standardized approach allows the ICT system's security engineer and CISSO of each organization within such a relationship to design the system or system component such that it maintains the requirements that keep it in compliance with the multiorganization ICT architecture and security strategies. Likewise, standardization assists in meeting the requirements for those controls that mandate the firewall installation (in this case) be integrated to the networks of supply chain organizations dependent on supplier or acquirer access and be upgraded on a regular schedule. Having a clear understanding of each supply chain organization's enterprise architecture and security strategies, security control implementation can be better managed to provide testing of leverage system security requirements and development and deployment of system services. Moreover, as the supply chain relationships continue to grow in terms of the level of security awareness, staff trained in identifying network/telecommunications threats and vulnerabilities can take appropriate measures to protect from and detect them. The net result is likely increased SCRM efficiency and substantial project cost savings.

The scope of implementation includes the activities that effectively allocate and integrate, to the specific ICT components, the controls identified in the Select Step of the RMF that will provide protection for the entire ICT architecture. Many security managers agree that control allocation is one of the most important and sometimes time-consuming activities of risk management. Successful implementation requires the coordination of each individual organization within the system supply chain offering common controls for inheritance purposes, in addition to each affected business unit and the ICT staff facilitating and supporting the system design and development. Put differently, it is through the identification of implementation requirements that organizations must clearly understand the supportive nature and implications of implemented controls within organizations throughout the entire supply chain, in order to provide an end-to-end architecture of system security and SCRM practices. This is not an easy task, but with the activities we discussed from the ISO/IEC 12207 Agreement process, the knowledge each organization requires for successful control allocation, and the contractual obligations set forth for the implementation of specific controls, it can be adequately documented and effectively communicated.

Consider the implementation of end-to-end SCRM controls; it would be important, for the sake of streamlined SCRM, for each organization within the supply chain to be consistent in the management controls that have been implemented. For example, CM-1 of NIST SP 800-161 states that each organization within the supply chain must define configuration management policy and procedures and that those policies and procedures should address the entire SDLC. The control guideline continues by stating that configuration management policy should take into consideration such implementation aspects as configuration items, data retention for configuration items, and tracking of the configuration item and its metadata. Further, the guideline emphasizes that each supply chain organization should coordinate with system integrators and external service providers regarding the configuration management policy. That last point is perhaps most difficult for configuration management and other management controls. Each organization is unique in terms of its enterprise infrastructure.

Therefore, the implementation of each management control is unique in adequately supporting that architecture. However, it is important to consider the means by which management controls have been implemented by third-party partners to prevent chaos within the decision managing processes of the organization's ICT management structure.

Aside from management controls, there are cases in which technical or operational controls are only implemented in specific components or subsystems and are not required across the entire supply chain. For instance, a system contained within one supply chain organization may have functional units that do not require data storage offsite, while another organization may have a portion of the ICT system requiring offsite storage for, among other purposes, compliance with legal and industry regulations. In these instances, the offsite data storage controls are considered separately and are not implemented in the part of the supply chain where the requirement does not exist. In general, the combination of each organization's underlying strategies, information processing needs, security requirements, system categorization, common control providers, and control allocation maintains a suitable balance among supply chain, system, and organization-provided security control measures. A clear understanding of each organization's strategies and controls supplied by common control providers helps to govern which of the required controls can be inherited by other supply chain organizations, which controls will be provided by the individual organizations, which controls will be implemented through a combination of the two, and which will be implemented as a hybrid control.

Security Control Documentation

Throughout the discussion of implementation, we have emphasized the necessity of structured life cycle processes and well-defined management controls that support those processes (in this case for the purpose of development of technical and operational security controls). Each of those processes must be supported by documentation that includes but is not limited to an assessment plan, a plan for process improvement, contingency plans, and project plans. Put more directly, each supply chain organization must develop (and when appropriate be able to distribute) a set of life cycle documentation that supports the development of required controls, validation (through traceability) of documentation that supports the premise that implemented controls meet established requirements, and the security plan updated with pertinent information about each implemented security control.

The documentation must also describe each control's categorization as common, hybrid, or system-specific and when appropriate be able to describe the extent by which each control is integrated with those of its supply chain partners. Beyond the scope of life cycle project plans, this documentation serves as the formal plan and explanation resource with information about the overall function and security implementation of each organization's system, including all required inputs and outputs. Additionally, it provides a snapshot of end-to-end security throughout the supply chain. Furthermore, the security control documentation defines the control's traceability to the control requirements as defined in NIST SP 800-161, NIST SP 800-53, NIST SP 800-53A, and required organizational or regulatory implications affecting facilities and how a control is implemented.

Through documentation, organizations are able to effectively create a balance between the level of effort necessary for the controls to be implemented, scope, and the impact that implementing each control will have on the organization's underlying business functions, strategies, mission, or operations and those of its supply chain partners. At a minimum, the documentation should provide a detailed explanation about the security control implementation process, required facilities, test procedures, and appropriate references to the bodies of evidence for common and hybrid controls. Detail in control design documentation should also provide adequate descriptions of planned inputs, expected behavior, and expected output from each control implemented.

Once complete, the documentation becomes part of the authorization package (we will discuss authorization later in this chapter); therefore, the security team, system engineers, and other pertinent ICT personnel must determine if each of the required security controls allocated as system specific or hybrid is appropriately implemented and adequately protects the system as designed and that the system life cycle documentation and requirements match the configuration of the system and its components across the supply chain. The most common approach to accomplishing this sense of certainty is executing traceability testing procedures designed to verify that the controls are documented and added to the system test processes. The key point is that, much like all forms of ICT system development, the end result of the implementation must be able to trace back to security requirements. Additionally, the authorization package requires an updated security plan. As you will see in our discussion of authorization, through sharing the documentation contained within that package, a streamlined SCRM process will manifest.

Supply Chain Security Control Assessment

FIPS 200 stipulates that organizations develop schedules for the assessment of security controls in organizational information systems as a way of measuring the effectiveness of each control within the context from which it was applied. The standard continues by stating that organizations must develop and implement action plans that correct discovered deficiencies and reduce or eliminate vulnerabilities in the organizations' ICT systems. The guidelines provided in the "Security Assessment and Authorization" family controls defined in NIST SP 800-161 put that requisite of assessment into the context of the supply chain by stating that organizations should integrate ICT supply chain implications into ongoing security assessment activities. Such implications include the organization's ICT systems and ICT supply chain infrastructure and external assessments of system integrators and external service providers. The guidelines emphasize that each supply chain organization should assess documentation and tracking of chain of custody and system integration between organizations, verify the existence of ICT supply chain security training, and verify each supplier's claim of conformance to security, product integrity, and the use of validation tools and techniques for noninvasive approaches to detect counterfeits or malware.

In general, the intention of assessment (as defined by the RMF) is that once security controls are implemented, they should be assessed to ensure that the organization and each organization within the supply chain achieves the desired level of effectiveness from those controls. More specifically, security control assessment is a process put into place by the organization in order to review the managerial, technical, and operational security controls that have been implemented into the ICT system and the organization's managerial structure. Such an assessment helps the organization determine if the controls were put into place correctly, operate as intended, and produce the desired outcomes as defined by the security requirements. To that extent, assessment activities go well beyond the degree from which verification and validation were performed during the implementation step.

Before expanding further, we should clarify the context from which the term "Risk Assessment" is used in NIST guidelines. Very often confusion exists regarding when risk assessment should be performed, before or after security control implementation, which creates a "which comes first the chicken or egg" phenomenon. The short answer is that it should be performed before *and* after. NIST SP 800-161 uses NIST SP 800-39, *Managing Information Security Risk: Organization, Mission, and Information System View,* to integrate ICT SCRM into risk management tiers and the risk management process. To a large extent, NIST SP 800-39 describes risk assessment from a risk analysis perspective, in which tasks are identified in order to assist an organization in identifying the most probable threats to an organization and analyze the related vulnerabilities of the organization to these threats. It is a very high-level overview of your technology, controls, and policies/procedures to identify gaps and areas of risk. The RMF presents risk assessment as the means for evaluating existing security and controls and assessing their adequacy relative to the potential threats of the organization. In short, both sets of processes should be in place to ensure SCRM is appropriate for adequately identifying and mitigating risk throughout the supply chain.

The Four Tasks of Security Control Assessment

The RMF identifies four tasks associated with assessment:

1. Develop, review, and approve a plan to assess the security controls
2. Assess the security controls in accordance with the assessment procedures defined in the security assessment plan
3. Prepare the security assessment report documenting the issues, findings, and recommendations from the security control assessment
4. Conduct initial remediation actions on security controls based on the findings and recommendations of the security assessment report and reassess remediated control(s), as appropriate

To facilitate the assessment tasks of the RMF, NIST SP 800-53A R4 *Assessing Security and Privacy Controls in Federal Information Systems and Organizations* can be used. This guideline provides a set of common assessment procedures for evaluating the effectiveness of the security controls organizations have implemented in addition to those implemented by supply chain partners. The guideline also provides guidance for building effective security assessment plans and managing assessment results. To support the effective assessment of technical controls, NIST has provided the guideline SP 800-115, *Technical Guide to Information Security Testing and Assessment*, which presents review, technical testing, and examination techniques that organizations can use to perform assessment for technical controls (Figure 4.4).

1. Develop, Review, and Approve a Plan
    During this activity of the assessment step, each supply chain organization must appoint an assessor whose first task is to develop an assessment plan to guide the assessment processes and procedures to follow. At a minimum, the plan should include which system components to assess, what automated

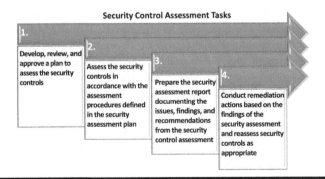

Figure 4.4   Security control assessment tasks.

tools or manual processes will be used, and a description of clearly defined assessment boundaries.

An accompanying test plan should identify agreed-upon rules of engagement (ROE) that have been approved by the information system owner, affected supply chain partners, and authorizing official. Too often, there is confusion amongst supply chain participants about the extent to which ROEs are applied. For streamlined assessment to be effective, it should include a definition of the scope and depth of the assessment, a point of contact (POC) within each organization for when unexpected events or incidents occur during testing, acceptable tools and techniques that must be used during assessment, and the appropriate levels of access to each system component necessary to complete the assessment or test. The test plan is based on a Body of Evidence (BOE) presented by the information system owner to the individual (whether within that organization or a representative from a supply chain partner). Put simply, the BOE comprises the contents of the system security plan, system architecture, and enterprise architecture documentation that includes associated policies, user guides, and previous test results. An accurate and thorough BOE ensures the capacity in which an assessor from any supply chain organization can understand the detail of how each interconnected supply chain system is built and the level of detail necessary to adequately test the systems security control implementation.

The first task in planning and preparing for assessment is to identify the controls that are to be assessed. In the case of supply chain relationships, the degree by which assessment is performed may extend well beyond a single organization's boundaries. NIST SP 800-53A contains assessment procedures for every control and control enhancement in the security control catalog of NIST SP 800-53, and it is echoed in NIST SP 800-161. The NIST catalogs are an excellent resource for beginning the decision-making process regarding which assessment procedures to follow. However, it is not uncommon for organizations to adapt proprietary or industry-based assessment procedures to achieve the intended assessment objectives, especially in supply chain relationships that, at times, include multiindustry implications. Nevertheless, regardless of the means by which assessment methods are chosen, the selection of those procedures must take into consideration criteria such as the impact level of each interconnected supply chain system and organization, in addition to assurance requirements of each organization that must be satisfied.

Once each organization has effectively established the scope of the assessment, other factors can be considered. At this point, a timeline for activities performed throughout the assessment can be established. Additionally, each organization can make the necessary decisions regarding the allocation of sufficient resources to the assessment process, including decisions related to how many assessors are necessary. When multiple assessors are assigned, it is important that each has sufficient expertise to evaluate assigned controls and that all assessors have a common understanding of what constitutes a "satisfied" finding.

As in the case of all ICT projects, legal implications must be considered before the organization begins implementing the assessment plan. For example, if an organization utilizes the services of external assessment organization, the legal departments of both organizations may be involved. Likewise, when assessment implications must be considered related to the interconnection between supply chain organizations, each organization must play a role in reviewing the assessment plan and providing specific clauses into contracts that dictate what can and cannot be done, relating to the assessments being performed. Confidentiality of information is also a concern. The legal departments may require external assessment organizations or supply chain partners to sign nondisclosure agreements that prohibit representative assessors from disclosing any sensitive, proprietary, or restricted information to unapproved parties. Within the agreement, privacy issues should be addressed. Each legal department should be aware of any privacy concerns that the organization may have and address potential privacy violations before the assessment begins. Finally, captured data may include sensitive attributes that do not belong to the organization or personal employee, data that may create privacy concerns. Each legal department has the responsibility to determine data handling requirements to ensure data confidentiality is intact.

2. Assess the Security Controls

Once the assessment plan has progressed through the appropriate approval process, management oversight of each supply chain organization must ensure that security control assessment proceeds according to the schedule and approach specified in the plan. One way to look at the activities associated with performing assessment is to think of it as building an *assurance case*, a term built from the work of the Software Engineering Institute of Carnegie Mellon University. An assurance case consists of evidence (obtained through performing activities of the SDLC) that the controls in the ICT system have been implemented correctly, operate as intended, and produce the desired outcome based on established security and privacy requirements. In doing so, the organization presents the evidence in a way that assists decision-makers in making effective risk-based decisions.

To be effective, the assessment should adequately verify the implementation of security controls documented in the system security plan and agreed-upon documentation set forth by the conglomeration of supply chain partners, by examining evidence produced through interviewing members of the security implementation team of each organization and testing the controls based on the criteria specified in each organization's assessment plan to validate that they function as expected and verify that evidence shows that the security controls continue to meet documented requirements.

The assessment process should follow the predetermined procedures for each control set forth in the security assessment plan through examining, interviewing, or testing applicable assessment objects and reviewing available evidence in order to make a determination for each assessment objective. The goal of the assessor is to find adequate evidence within each assessment procedure to render a result in a finding of "satisfied" or "other than satisfied."

Regardless of the procedure, it is important that security control assessment findings be objective, evidence-based indications of the way the organization has implemented each security control. Since documentation and observation are generally used as a source of evidence for assessed controls and are frequently distributed to supply chain partners, such evidence must be correct and complete and present a level of quality that provides its own evidence of accuracy. Moreover, documentation of security control assessment results should be presented at a level of detail appropriate for the type of assessment being performed and include required criteria-consistent organizational policy and agreements set forth through supplier–acquirer contracts.

3. Prepare the Security Assessment Report

At the conclusion of the formal assessment of security controls, each organization should prepare a draft security assessment report. Included within the report are the assessment findings and indications of the effectiveness determined for each security control implemented for the ICT system. For ease in creating the report and ensuring that it contains the correct content, we recommend the general format provided by NIST in Special Publication 800-53A. NIST suggests that the results of security and privacy control assessment directly impact the way controls are implemented across the entire supply chain. Further, the guideline emphasizes that the assessment has an influence on what is contained within the security and privacy plans and other plans of action and milestones that directly impact the security posture of the organization and those supply chain partners that depend on such documentation in order to provide the end-to-end implementation and assessment strategies that contribute to a unified SCRM effort.

Once prepared, organizational management along with the ICT system users, affected supply chain partners, and common control providers review the security assessment reports, privacy assessment reports, and updated risk assessment to determine the next steps required in response to the identified weaknesses and deficiencies. In doing so, NIST recommends that the assessment report include the labels S for satisfied and O for other than satisfied in providing visibility into specific weaknesses and deficiencies of security or privacy controls that have been identified within the ICT system or other influential system within the supply chain. Additionally, the report should document assessment findings and provide recommendations for correcting the identified control weaknesses, deficiencies, or other-than-satisfied determinations made during the assessment. Specifically, NIST SP 800-53A stipulates the following content be included within the report:

- The information system name
- The impact level assigned to the system
- Results of previous assessments or other related documentation
- The identifier of each control or control enhancement assessed
- The assessment methods and objects used and level of depth and coverage for each control or enhancement
- A summary of assessment findings
- Assessor comments or recommendations

It is not uncommon for the assessment team to provide its assessment results in an initial security assessment report as a means to communicate missing evidence or provide corrective actions for identified control weaknesses or deficiencies before the security assessment report delivered in final draft form. Likewise, it is common for the assessment team to reevaluate any security controls added or revised during this process, which includes the updated assessment findings in the final security assessment report.

4. Conduct Initial Remediation Actions

The final draft of the security assessment report provides awareness each supply chain organization needs into specific weaknesses and deficiencies in the security controls within its own ICT infrastructure or partnering organizations. While it might seem impractical for supply chain organizations to voluntarily share assessment findings, accessibility of the assessment report can and should be included within the contractual agreements signed by each organization at the outset of the supply chain relationship. The findings generated during the security control assessment can be thought of as a disciplined and structured approach to mitigating risks according to the priorities set forth by each organization. The NIST RMF stipulates that once the final assessment report is completed, each organization should use the security assessment report to develop a plan to resolve (or remedy) those security control weaknesses and deficiencies discovered through the assessment process. During this part of the assessment process, organization officials and control providers engage in discussions about the report and work collaboratively to make decisions related to "next steps" for improvement. This collaborative engagement of discussion may result in a decision that certain findings are frivolous and realistically present no significant risk to the organization.

Alternatively, it may be determined that specific findings within the report are substantial enough to require immediate remediation actions. In some cases, security controls that have been identified as weak or deficient may be so significant to the vulnerability of the system that remediation is necessary prior to the system's moving into production. If such a significant vulnerability exists and is outside the control of the organization, actions should be taken to work collaboratively with the supply chain partner from which the vulnerability is established. If a solution cannot be realized, a "last resort" alternative would be to sever the relationship with that partnering organization in accordance with the contractual agreements.

When controls that have been identified as a weakness or deficiency have been corrected, they must be tested and reassessed, ensuring that all corrective actions are compliant with the organization's Configuration Management policy and have gained appropriate approval through the Configuration Control Board, in accordance with the managerial controls CM-1 and CM-3 as defined in NIST SP 800-161. To that extent, actions implemented to mitigate risk are implemented and in turn verified as a means for ensuring that the implementation process was completed accurately.

Verification is the process in which the organization conducts an audit of the system consistent with the management controls of the NIST SP 800-161 Audit and Accountability family. Through this process, the system is retested against predetermined test cases and documentation prepared as a means for holding the individuals performing the verification processes accountable. The advantage of a system audit is that it includes details of technical verification of the changes that have been implemented in the system and can be conducted by internal security personnel or an external security test organization. The audit team should use a defined mitigation strategy as a means of ensuring that each action is completed.

Most organizations choose to adopt an issue resolution process designed to assist in determining the appropriate actions to take with regard to the security control weaknesses and deficiencies identified during the assessment. The practice of utilizing issue resolution can help address vulnerabilities and associated risk, false positives, and other factors that may provide useful information to the organization regarding the overall state of security of the ICT system across the supply chain including system-specific, hybrid, and common control effectiveness. Furthermore, the issue resolution process provides valuable assistance in ensuring that only substantive items are identified and transferred to the plan of actions and milestones.

Implications of Security Control Authorization to the Supply Chain

The authorization step of the Risk Management Framework is often misinterpreted as having few, if any, implications to the underlying practices of SCRM; this is not an accurate interpretation. To the contrary, it includes the documentation of the acceptance of a formally sanctioned, organization and supply chain wide systematic approach to the risk management needs of a given ICT infrastructure.

The management of risk is a complex and multilayered process that requires top-to-bottom involvement of not only one organization but all organizations with which it associates. As we have already mentioned, risk management is intended to leverage trust and confidence for any ICT system across the entire spectrum of the organization and the cultures on which it bases its decision-making. We remind you that trust is a vital element in the consideration of good risk management practice. The means by which an organization ensures trust will influence its long-term corporate relationships as well as the internal and external aspects of doing business.

The underlying practices of authorization involve formal certification and accreditation (C&A) processes. However, there are numerous approaches to C&A. The Federal Government's FISMA provides a clearly defined set of requirements that can be satisfied by adopting the recommendations of NIST SP 800-37 R1, *Guide for Applying the Risk Management Framework to Federal Information Systems: A Security Life Cycle Approach*. The guideline provides details for creating and distributing a FISMA security authorization package; which includes the security plan, security assessment report, and plan of action and milestones (all documents of value to each organization participating in a given supply chain relationship). The guideline provides a general explanation of the necessary mechanisms for establishing the exact criteria to be included in the security plan as well as how an action plan and milestones can be created to ensure practical direction for all supply chain partners. Further, the guideline explains how the accreditation plan documents the organization's specific approach and strategy for finding and remediating a particular security weakness or operating deficiency that has been identified through security control assessment.

In general, C&A provide the mechanisms necessary for establishing a well-defined approach to evaluating, describing, testing, and authorizing systems and their associated activities prior to or after a system is put into operation. Certification simply entails a formal process for confirming a set of characteristics of a system component, developer or system user, or organization. This process is often provided by some form of external review, education, assessment, or audit. Recall from our discussion of the RMF assessment step that NIST SP 800-161 provides criteria for 16 audit management controls that, when implemented, will provide mechanisms to simplify the certification process across the supply chain. Accreditation is a formal organizational process for performing certification. It is the step that each supply chain organization must take in order to certify its competency, authority, or credibility. In most cases, this process is provided by a third-party accrediting institution often known as a "certifier." Accreditation ensures that the testing and audit practices of the certifier are sufficient to distinguish conformance with a given standard, or regulation, as well as that the audited parties behave ethically and employ appropriate control assurance.

The Four Tasks of Security Control Authorization

NIST SP 800-161 (CA-6) stipulates that all supply chain organizations have security authorization in place and identifies it as a high-priority management control. The catalog describes the control as:

> official management decisions, conveyed through authorization decision documents, by senior organizational officials or executives (i.e., authorizing officials) to authorize operation of information systems and to explicitly accept the risk to organizational operations and assets, individuals, other organizations, and the Nation based on the implementation of agreed-upon security controls.
>
> *(Boyens et al., 2015)*

The control guidelines emphasize that it is the responsibility of the authorizing officials to be accountable for security risks associated with the operation and use of the ICT systems. In essence, due to the level of authority given to the authorizing officials, they understand and accept information security-related risks on behalf of the organization. To assist authorization officials in performing their duties, the NIST RMF outlines four core authorization tasks each supply chain organization must perform; they are summarized in Figure 4.5 and as follows:

1. Prepare the plan of action and milestones based on the findings and recommendations of the security assessment report.
2. Assemble the security authorization package and submit the package to the authorizing official for approval.
3. Determine the risk to organizational operations, organizational assets, individuals, and other organizations.
4. Determine whether the risk to organizational operations, organizational assets, individuals, and other organizations is acceptable.

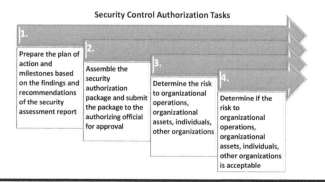

Figure 4.5   Security control authorization tasks.

1. Prepare the POA&M.

   Management control CA-5 of NIST SP 800-161 stipulates that organizations participating in supply chain relationships develop a plan of action and milestones (POA&M) as a way of documenting the planned remedial actions to correct weaknesses and deficiencies within the ICT system, identified during assessment and documented in the assessment report. The control guideline continues by stating that the organization must have processes in place to update the existing POA&M as a result of findings generated from assessments performed on security controls, security impact analyses, and continuous monitoring activities.

   The POA&M provides the authorizing official the vital risk impact information needed to make authorization decisions. The authorizing official must be able to consider many factors when making risk acceptance decisions, including the impact of an ICT system on business goals and objectives, impact of the system on the system objectives of supply chain partners, the effect such decisions have on corporate reputation, and the operational elements such as information assets and people that fall within the system boundary. This is not a "one size fits all" decision process. It requires the ability to balance mitigation of all of the known risk factors against the efficient and effective operation of the business; such information is provided within the POA&M.

   The security assessment report contains the detailed findings from the testing and evaluations of each security control. Further, it identifies which of those findings could be considered as acceptable risk and which are not acceptable. Unacceptable (or residual) risks are findings that are considered detrimental for the operation of the system. Therefore, the organization must have a plan for implementing solutions and mitigating risks. It is the responsibility of the information system owner to prepare the POA&M for remediation and mitigation. The document is then submitted to the authorizing official, and it becomes one of the three documents contained within the authorization package. As noted previously, the plan is intended to describe the tasks that are planned to remediate any weaknesses or deficiencies in the security controls identified during the assessment. Additionally, the plan must describe the strategy for risk acceptance in order to effectively address system vulnerabilities.

   In essence, the POA&M provides the means by which the authorizing official can monitor the progress in remediation, or rework, for each weakness or deficiency identified during assessment and provides a reasonable mechanism for managing an audit trail of risk mitigation. To that extent, the plan and the associated milestones can be used by the authorizing official to monitor the organization's progress in correcting weaknesses or deficiencies noted during the security control assessment.

   The details contained within the POA&M theoretically drive the successes or lack thereof in the authorization process. It specifies the actions that will be taken and the milestones that will be met to ensure proper certification of system capabilities. This document provides a precise set of activities that will be carried out along with a set of tailored recommendations for how all necessary remediation, rework, or additional development will be carried out prior to implementation.

2. Assemble and Submit the Security Authorization Package.

Upon completion and agreement of the POA&M, focus then turns to the assembly and submission of the authorization package to the authorization official for formal acceptance. The information system owner is responsible for making sure all components are included and delivery to the authorizing official has taken place. The authorization package should include the security plan, the security assessment report, the POA&M intended to address any identified weaknesses or deficiencies. The information in these key documents is then used by authorizing officials to make a risk-based authorization decision.

In order to provide the necessary mechanisms to maintain and update control status information for authorizing officials, organizations will often utilize automated tools to prepare and manage the content of the package. Automated tools provide an orderly, disciplined, and timely way to update the security plan, the security assessment report, or the POA&M. Likewise, they provide the ability to achieve near real-time risk management and capabilities for ongoing authorization for each system within the supply chain. That said, it becomes clear that automation facilitates more cost-effective and meaningful reauthorization processes.

3. Determine the Risk.

Through risk determination, the organization must effectively assess the risk to the organization through threats and vulnerabilities to organizational assets, personnel, or other supply chain organizations. This is accomplished by each organization following a defined process in which it determines ICT supply chain risk by considering the chance that known threats could exploit vulnerabilities throughout the ICT supply chain and the result in consequences that harm the ICT system or prevent the organization from achieving its organizational goals and objectives. Organizations use threat and vulnerability information with likelihood and consequences/impact information to determine ICT SCRM risk either qualitatively or quantitatively.

The context in which we state "the chance that known threats could exploit vulnerabilities throughout the ICT supply chain" requires further elaboration. Determining that known threats can exploit vulnerabilities requires putting into perspective knowledge of the threat sources, the identified vulnerabilities, and those supply chain organizations that would be affected by the existence of the vulnerability or threat. Serious consideration must be given to the extent by which the exploitation impacts each supply chain organization's mission. ICT supply chain risk assessment should, at a minimum, consider two criteria:

– The possibility that the supply chain is capable of being compromised
– The possibility of a system or component of the supply chain being compromised

Further, during assessment, consideration must be given to such factors as:

- The type of threats to which a system or system component may be susceptible
- An understanding of who the supply chain organization is protecting itself from and the intentions and tools being used to perform the attack
- The degree of exposure each system component has to external access
- Analysis of known system, process, or component vulnerabilities supported by data in order to determine the likelihood of future ICT supply chain compromise

Through risk determination, the objective is for each organization to determine the ease in which a system or component can be compromised and its ability to detect the method used to introduce or trigger a vulnerability. The goal is to understand the net effect of each identified vulnerability, which includes the threat information to determine the likelihood of successful attacks. Collected data becomes useful in providing statistical analysis toward conclusions about whether an attack will occur and accessibility characteristics within the organization and throughout the supply chain.

4. Determine Whether the Risk Is Acceptable.

The final task of authorization is the acceptance of residual risks that result from performing activities associated with risk assessment. To make accurate acceptance decisions requires the involvement of the organization's management team and often includes collaboration with management from supply chain partners. To that extent, risk acceptance requires the communication of residual risks identified through the risk assessment process and detailed within the risk assessment report.

Put simply, the decision of risk acceptance is based on criticality. In other words, the centrality and sensitivity of the ICT system being considered for authorization largely determines the amount and degree of risk management control acceptable for that system. The degree of confidence in the risk acceptance decision is a reflection of the amount of rigor evident within the process of risk analysis. That level of rigor is then factored into the assessed level of vulnerability or criticality of the ICT system and the known threats associated with each of these individual risks.

Once accepted, residual risks are considered as risks the management of the organization knowingly takes. The level and extent of accepted risks comprise one of the major parameters of the SCRM process. In other words, the higher the accepted residual risks, the less the work involved in managing risks. This does not mean, however, that once accepted the risks will not change as a result of future SCRM activities. Rather, as future risk management tasks are performed, the severity of these risks will be measured over time. In the event that new assertions are made or changing technical conditions identified, risks that have been accepted need to be reconsidered. Finally, a formal statement of risk tolerance is prepared with a statement of assessed criticality, risk tolerances, and concomitant risk controls. This is distributed along with a plan for monitoring each mitigated risk over time.

Supply Chain Risk Continuous Monitoring

If you are reading this book, you may have interest in or work in some facet of the SDLC. Therefore, you are likely familiar with the final stage of the life cycle, maintenance. During the maintenance phase, organizations are faced with the task of understanding the role of the system under the assumption that it was "built right," identifying flaws of the system, and making the necessary system changes in order to correct those flaws or allowing the system to support the organization, as new strategies, objectives, and missions are developed. The fact of the matter is that systems and environments change over time. Thus, in the context of SCRM, there is always a need to ensure that a suitable end-to-end security response continues to be maintained for each ICT system or system component characterized as a specific threat environment. To achieve an adequate level of security response, a formal control monitoring process is needed and must be capable of continuous assurance of the appropriateness and sufficiency of the control response within the known threat environment and in accordance with any documented risk acceptance decisions.

Recall from our previous discussions of NIST SP 800-161 in previous chapters that the standard provides a clear definition, from the perspective of the three-tiered organizational structure defined in NIST SP 800-39, of the monitoring responsibilities that each supply chain partner must perform in the presence of an underlying implementation of the SCRM process. Step six of that process, *monitor*, specifies that the objectives within this phase include:

■ *At the Organization Tier*
  – Integrate SCRM into the existing continuous monitoring program
  – Monitor and evaluate constraints and risks for change and their impact at the enterprise level
  – Monitor the effectiveness of risk response at the enterprise level
■ *At the Mission/Business Process Tier*
  – Identify which business functions need to be monitored for supply chain change and assessed for impact
  – Integrate SCRM into continuous monitoring processes
  – Monitor and evaluate constraints and risks for change and their impact at the business process level
  – Monitor the effectiveness of risk response at the enterprise level
■ *At the System Tier*
  – Monitor the system level requirements' response to change and assess their impact
  – Monitor the effectiveness of system-level risk response

In short, the monitoring objective identified at each tier represents the need for analysis of individual occurrences in the organization, business process, and system environment that might pose a threat at that level. The status of existing controls is evaluated with regard to its ability to protect the organization or supply chain partners from newly identified threats. The process itself is activated when an event occurs that may impact the overall organization, mission, or system security status. From the analysis that gets performed, the relevant organizational stakeholder or manager is given a set of recommendations that might include options such as *change* or *patch*. They could recommend seeking insurance from a third party; they might also simply recommend *accept*.

In order to maintain a sufficient understanding of the risk at each tier, each supply chain organization must institute a properly targeted risk-monitoring function. The outcome of that function should be the results of ongoing qualitative and quantitative analyses of any newly identified or emerging risk event. In addition to newly identified risks, that risk-monitoring function should have the capability to perform the analyses required in order to confirm that currently existing risks are fully characterized and contained. Ideally, the execution of the risk-monitoring process should produce a continuous certainty throughout the supply chain, that the risks each organization considers priorities are understood and mitigated, and that any emerging risks are identified, analyzed, and appropriately dealt with, according to the processes of the risk-monitoring function, as they occur (Figure 4.6).

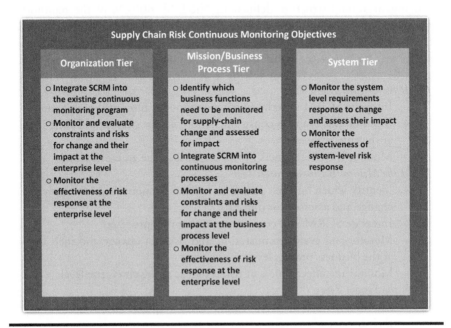

Figure 4.6 Supply chain risk continuous monitoring objectives.

The former point of identifying risks as they occur is a task easier said than done. Remember that the underlying purpose of risk monitoring is to continuously establish and maintain an appropriate set of managerial, operational, and technical controls that reduce or eliminate risk. For this reason, risk assessments are a critical component to the monitoring process. As we discussed earlier in the chapter, risk assessments ensure effectiveness at all levels of each supply chain organization by their ability to identify the specific threats to each organization, determine how likely those threats are to occur, and substantiate the consequences of each threat. Correctly done, the existing threat environment is periodically assessed to ensure that the current risk mitigation scheme is relevant and maintains its effectiveness.

The Seven Tasks of Security Continuous Monitoring

While NIST SP 800-161 and NIST SP 800-39 identify the tasks of continuous monitoring through activities performed at each organizational tier, the NIST RMF outlines seven core authorization tasks that each supply chain organization must perform and are summarized as follows (Figure 4.7):

1. Determine the security impact of changes to the ICT system and its environment.
2. Assess selected technical, management, and operational security controls.
3. Conduct remediation actions resulting from the results of monitoring activities, risk assessment, and items in the POA&M.
4. Update the security plan, security assessment report, and POA&M.

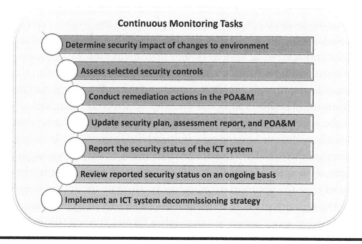

Figure 4.7   Continuous monitoring tasks.

5. Report the security status of the ICT system to the authorizing official and other appropriate organizational officials on an ongoing basis.
6. Review the reported security status of the ICT system on an ongoing basis to determine whether the risk to organizational operations, organizational assets, individuals, or other organizations is acceptable.
7. Implement an ICT system decommissioning strategy, when needed.

Determine the Security Impact of Changes

Every day ICT systems change in one way or another. New components are added; scheduled upgrades are performed to existing hardware, software, or telecommunications; and modifications are made to the physical environments from which the systems operate. The RMF stipulates that each organization implement a process with adequate management oversight to allow for the controlling and documenting of changes to the ICT system and its operating environment. Specifically, the organization must implement each of the 11 controls of the configuration management family listed in NIST SP 800-161 in order to support continuous monitoring activities. The controls should allow for the recording of any specific changes that occur to the systems hardware, software, telecommunications, or the environment from which they operate. While such information is valuable for the common practices of configuration management, such as version control or documentation of updates, it also provides the information necessary for proper security implementation and allows the organization to track changes that directly affect the underlying risk management strategy. The information is also useful to the organization common control providers and supply chain partners as a means for assessing the potential security impact of the documented changes.

Through security impact analysis (CM-4 in the NIST SP 800-161 control catalog) the organization analyzes changes to the information system to determine potential security impacts prior to change implementation. Security impact analysis may include, for example, reviewing security plans to understand security control requirements and reviewing system design documentation to understand control implementation and how specific changes might affect the controls. This analysis may also include assessments of risk to better understand the impact of the changes and to determine whether additional security controls are required. Intentional changes to the system or its environment may indirectly affect the security controls currently in place, thus producing new vulnerabilities that must be mitigated.

If the results of the security impact analysis indicate that the changes can affect or have affected the state of security in the system or its environment, the organization should initiate the appropriate processes to alleviate the security repercussions and update the security plan, assessment plan, or POA&M as necessary. However, it is important that any changes go through a formal configuration management process supported by the organization's management, Configuration Control Board, and Chief Information System Security Officer. When changes affect the implementation of security controls or the ability to perform business functions of supply chain partners, those organizations must be notified as well. Once the changes have been implemented, it is the responsibility of the authorizing official to review the updated security assessment report to determine if a formal reauthorization is warranted. It is not uncommon for routine changes to an ICT system or its environment to be handled directly through the organization's continuous monitoring program, which provides the mechanisms for ongoing authorization and near real-time risk management.

Assess Selected Security Controls

During initial security authorization, each security control is assessed for effectiveness. In much the same way that systems are continuously improved during the maintenance phase of the SDLC, the organization performs routine assessments of subsets of management, operational, and technical controls during continuous monitoring. Considering the implications of supply chain, it is important to understand that such controls may exist within an organization or be managed by supply chain partners. In some cases, the agreement made between a supplier and acquirer will specify the control subset definitions and assessment schedules, while other agreements include the subset definitions and predetermined third party with the responsibility of assessing the controls, allowing for neutral views of the controls and their security impact within a given system or organization. Assessor independence during continuous monitoring, whether for a single organization or supply chain relationship, promotes efficiency of evaluation and allows for reuse of assessment results when reauthorization is required.

The predetermined selection of security controls subsets to monitor and the frequency of monitoring are based on the monitoring strategy of each organization and largely dependent on former assessment results. NIST recommends that the selection be drawn upon by any of the following:

- Security control assessments conducted during authorization
- Observations made through the employment of continuous monitoring activities
- Testing and evaluation of the ICT system as part of the SDLC or audit processes

Conduct Remediation Actions

Upon completion of the assessment of security control subsets, pertinent assessment information produced during the assessment should be provided to the information system owner, common control provider, and affected supply chain partners in an updated security assessment report. In turn, information system owner, common control provider, and supply chain partner begin the process of remediation on any items listed as incomplete in the POA&M and make the necessary corrective actions resulting from findings produced during the monitoring of security controls. As organizations begin to implement remedial actions, care should be taken to include any recommendations provided by the security control assessor. As each security control is modified, enhanced, or added during continuous monitoring, it must be reassessed to ensure that appropriate corrective actions are taken to eliminate vulnerabilities or mitigate the identified risk.

Update the Security Plan, Security Assessment Report, and POA&M

As a way to promote end-to-end SCRM, it is necessary for each organization to update and distribute to appropriate parties the security plan, security assessment report, and POA&M on a regular basis. It is important that the updated security plan reflect any changes to security controls based on assessment and remediation actions taken during the continuous monitoring step. Likewise, the updated security assessment report should clearly delineate the assessment results stemming from the evaluation of control subsets from assessment activities performed to measure control effectiveness. Finally, the updated POA&M should adequately:

- Provide details of progress to items listed in the plan
- Provide the necessary details related to vulnerabilities identified through security impact analysis and/or security control monitoring
- Provide details into how the vulnerabilities will be mitigated

As we have emphasized throughout this book, a key component of successful SCRM is the inclusive nature of end-to-end security awareness. The information provided by these three documents helps to ensure adequate awareness of the security posture of the ICT system within each supply chain organization and the management oversight in place to provide the necessary support.

Most organizations will have their own schedule from which the three plans get updated. However, in a supply chain relationship, there is a definite dependency by supply chain partners to have the updates available. Update schedules can easily be negotiated as part of the Agreement process between supply chain partners and a clause included within the contract between the organizations. It is imperative for each partner to provide accurate and timely information regarding each organization's security posture since the information provided affects decisions made by authorizing officials and other management teams within each supply chain organization.

Report the Security Status

Reporting the status of activities performed in continuous monitoring to the authorizing official is a critical component because it keeps the official informed of the security status of the system and gives the information needed to make decisions about the ongoing status of the system's authorization to operate. This type of reporting can be event-driven or timed. Event-driven updates are triggered by some action such as a system breach or newly identified vulnerability. Time-driven updates are performed on a predetermined basis: weekly, monthly, or quarterly. In some cases, updates are triggered by both. This is particularly true when considering supply chain relationships, in which an agreement has been made regarding the timeframe and/or when a viable vulnerability has been identified. It is not only important for the authorizing official from the organization from which the vulnerability has been identified, but also such information is valuable to supply chain partners in order to measure the extent of exposure their systems have to the newly identified security threat.

When not bound by contractual agreements of supply chain, each organization has the flexibility to determine the format of the reporting process. However, some general guidelines include the current security state of each security control, the status of the continuous monitoring process, details related to any updated controls, details of newly identified vulnerabilities, mitigations in place to circumvent the newly identified vulnerabilities, and changes to the system's physical environment. When reviewing the system updates, the authorizing official will use this information to ensure that the system is not creating undue risk to the organization that would otherwise require the system to undergo reauthorization. In short, the system's continuous monitoring reporting program should be designed such that security controls can be evaluated efficiently and effectively as a driver for achieving the system's plans and the requirements set forth by the organization and requirements of supply chain partners.

Review the Reported Security Status on an Ongoing Basis

The main objective of this task is to provide ongoing risk determination and acceptance. This is accomplished through a review of the security status documentation by the authorizing official on an ongoing basis. This evaluation not only legitimizes that the changes made to the system increase system risk, but also determines the extent of risk imposed to the organization or supply chain partners. Once a decision is made on how the changes impact the authorization status of the system, the authorization official must prepare the necessary documentation aimed at communicating to the information system owner the security/authorization status of the system. If there are new limitations to the system that result from the changes, documentation is prepared to communicate those limitations to appropriate individuals within the organization and when necessary the affected supply chain partner.

Through this process, the system owner is able to ensure that the system is implemented with an adequate level of security. Further, it ensures that risk determination and acceptance within the system and across all organizations of the supply chain stay at an appropriate level. Once complete, the updated risk determination and risk acceptance documentation is forwarded for inclusion in the organization's risk management plan.

Implement an ICT System Decommissioning Strategy

At the end of a system's useful life, it must follow a process for removing it from service (or, decommissioning). This process becomes even more challenging when supply chain relationships have been built around the use of that system. Each organization must have an implemented system decommissioning/disposal strategy that takes risk management into consideration. First, the organization needs to ensure that all security controls addressing information system removal and decommissioning are implemented. For example, management control SA-19 "Component authenticity" of NIST SP 800-161 stipulates that the organization must dispose of components in such a way that the organization, its mission, operational information, and supply chain cannot be compromised. Other controls exist throughout the catalog, such as media sanitation, that ensure the organization is not putting itself or the supply chain partner at additional risk during the disposal process.

Second, organizational tracking systems that have been put into place through configuration management activities should be updated to identify the system or system component(s) taken out of service. Note that these records may not exist within just one organization. The updating process must propagate throughout the entire supply chain in order to generate end-to-end consistency of system information. The most effective way to establish the necessary organization-to-organization communication is through the use of security status reports, which reflect the new status of the ICT system. An adequate level of consistent system information across the supply chain will ensure implemented security assessment procedures can be followed to protect each organization against future risks and vulnerabilities.

Chapter Summary

By this point in your reading of this book, it should be evident that all systems have a supply chain of some sort. Putting systems into the context of their many components, each individual piece of the system has a supplier to one extent or another. When thinking in terms of software, each line of code came from somewhere, was tested by someone, and was packaged together before you brought it into your organization. If you're not thinking about your systems in terms of supply chain, that is not uncommon. Of course, your organization is also probably vulnerable as a result. That is not uncommon either. Regrettably, it is too often the case that a problem upstream in the system supply chain is disregarded by eventual end users; this vulnerability could expose all manner of private information across a network.

Many organizations upstream in a supply chain face serious risks because of an error earlier in the system supply chain.

Understanding the system supply chain will not eliminate errors in implementation. However, realization of its existence allows you to better manage risk. After all, cybersecurity is never about absolute risk elimination. It is about risk management. Understanding where that risk comes from means understanding where each of your system components has originated.

The hardware, software, and telecommunications that organizations use are composed of a multitude of parts that have been created, integrated, and assembled from multiple vendors. These vendors may be open source in nature, located overseas, or from unknown suppliers such as individual programmers operating out of their garages. In addition, contractors may have been used to create or purchase software modules. Previously written software may have been reused or custom software development may have occurred. Taken together, all these items create a system supply chain.

Organizations are increasingly acquiring COTS and open-source system products, all of which must be secure. At minimum, the system component should run as intended, produce results as expected, and resist attempts at compromise. In addition, any personal or organizational intellectual property data that systems utilize or access must be protected. Unfortunately, current approaches to the system and software Acquisition process do not account for the risk management issues of complex system supply chains and do not provide sufficient protections.

It becomes a matter of blind trust to assume that the vendor has implemented adequate cybersecurity controls along the entire supply chain and has performed adequate security testing against inherent software defects. In cybersecurity, however, blindness can lead to disaster. System defects in COTS can provide new avenues for attackers, which in turn can lead to a number of unanticipated issues for unsuspecting users.

DHS has long been collaborating with the NIST, international standards organizations, and tool vendors to create standards, metrics, and certification mechanisms from which tools can be qualified for system security verification. These efforts are helping to mature a SCRM approach.

As you have learned in previous chapters of this book, the SCRM process considers all sources of software, hardware, and telecommunication while identifying and quantifying risk from various perspectives. This risk can take many forms, from technical (intentional injection of malware into downstream software applications) to business (a hardware company going out of business and leaving no viable maintenance for products). A comprehensive SCRM process will look at people involved, processes in use, implemented technologies, and the entire Acquisition process to ensure that reliable suppliers are utilized at each step. Further, we should note that these risks continue to proliferate even after supplied products or services are placed into production. New threats emerge, and new attack patterns come into use that may impact previously secure system components. Understanding the entire supply chain and the risks involved at each stage helps to mitigate newly uncovered issues.

NIST is the standards creating organization for the U.S., and its Executive Order 13636—*Improving Critical Infrastructure Cybersecurity*—is one of the driving documents impacting the acquisition of cyber services. In this document, the President directs more information sharing across public and private organizations as well as operators of critical infrastructure to implement risk-based standards. This in part led to the creation of NIST SP 800-53, which defines Security and Privacy Controls for Federal Information Systems and Organizations. This document (along with SP 800-37) defines the RMF and six process steps:

- Categorize
- Select
- Implement
- Assess
- Authorize
- Continuous monitoring

This process has filtered into Federal acquisitions as a desire to address cyber concerns earlier in the procurement life cycle and system design. Federal acquisitions understand that the tech refresh cycle is substantially faster than the contract procurement cycle, which means that cyber protection requirements can quickly become outdated. Hence, a concept of "cyber resilience" is necessary to procure engineered systems that are able to survive an evolving and maturing cyber threat over time.

Vulnerabilities in systems expose each user to a tremendous amount of potential threats. By defining terms, establishing best practices, creating tools to test system components in both static and dynamic forms of operation, and identifying common system weaknesses, we can build and operate more securely. DHS has spearheaded a wide range of initiatives to further promote these goals and objectives, while NIST's RMF creates a standardized method to evaluate risks throughout the entire system supply chain. Still, even with our best efforts, some level of residual risk will remain in the system supply chain process. The goal is to minimize avoidable risk to acceptable levels and maintain a high level of confidence in the complexity of ICT systems in use around us.

Key Terms

accreditation: the formal attestation that all requirements and criteria have been met

authorization package: specific documentation collected during the Categorize, Select, Implement, and Assess steps of the RMF, evaluated as a means for authorizing the security controls of an ICT system

certification: formal documentation that an object under evaluation has met requirements

configuration management: a formal process to ensure the continuing of a logically related array of ICT components; the detailed recording and updating of information that describes an organization's hardware and software

continuous monitoring: a defined security process that enables information security professionals and others to see a continuous stream of near real-time snapshots of the state of risk to their security, data, the network, end points, and even cloud devices and applications

control assessment: the process of testing and/or evaluation of the management, operational, and technical security controls in an ICT system in order to determine the degree to which the controls are implemented correctly, operating as expected, and producing the desired outcome based on the security requirements of the system

risk assessment: estimate of likelihood and impact of all known threats; drives risk tolerance decisions

risk mitigation: the explicitly designed control for a given organizational threat

security control assessment plan: a set of predetermined activities and tasks that provide details of how a security control assessment will be performed. Some of the criteria of the plan include assessor identification, timelines for completing the assessment, controls to be assessed, and methods to be used in assessing each control

security control assessment report: provides the specific details of the controls assessed, methods used, and the findings, and conclusions made during the assessment process

security plan: a formal plan that provides a systematic approach and controls necessary to protect an ICT system from security threats and other forms of exploitation

validation: testing to ensure that the developed product provides the intended functionality

verification: the process of testing documented ICT requirements, to ensure that they have been met

References

Boyens, J., Paulsen, C., Moorthy, R., and Bartol, N., *NIST SP 800-161: Supply Chain Risk Management Practices for Federal Information Systems and Organizations.* Guideline, National Institute of Standards and Technology, Gaithersburg, MD, 2015.

Dempsey, K., Chawla, N. S., Johnson, A., Johnston, R., Jones, A. C., Orebaugh, A., Scholl, M., and Stine, K., *NIST SP 800-137: Information Security Continuous Monitoring (ISCM) for Federal Information Systems and Organizations.* Guideline, National Institute for Standards and Technology, Gaithersburg, MD, 2011.

NIST, Categorize step—Tips and techniques for organizations. *Computer Security Division-Computer Security Resource Center*, January 27, 2009, Accessed November 27, 2016. http://csrc.nist.gov/groups/SMA/fisma/Risk-Management-Framework/categorize/ QSG_categorize_tips-and-techniques-for-organizations.pdf.

NIST, *NIST SP 800 53a Rev 4: Assessing Security and Privacy Controls in Federal Information Systems and Organizations-Building Effective Assessment Plans: NIST SP 800-53Ar4*. Standard, NIST, Gaithersburg, MD, 2014.

NIST, Select step—Tips and techniques for systems. *Computer Security Division-Computer Security Resource Center*, January 18, 2011, Accessed December 3, 2016. http://csrc. nist.gov/groups/SMA/fisma/Risk-Management-Framework/select/qsg_select_tips- and-techniques-for-systems.pdf.

Chapter 5

Establishing a Substantive Control Process

At the conclusion of the chapter, the reader will understand the following:

- The specific purpose of National Institute of Standards and Technology (NIST) Special Publication 800-161
- The 21 principles for supply chain risk management (SCRM)
- How NIST SP 800-53(4) can help shape supply chain controls
- The use of standard regulatory requirements in securing supply chains
- The Control process for substantive SCRM

Introduction: Using Formal Models to Build Practical Processes

This chapter discusses the mechanisms for establishing an SCRM capability within any organization. The goal in each of these cases is to formulate a practical way for suppliers, acquirers, and security professionals to implement a unified process that will allow them to make informed decisions about the trustworthiness of sourced products. Information and communication technology (ICT) product supply chains are comprised of organizations, people, activities, information, and resources that move a product from bid to customer delivery. The problem is that supply chains are complex, and their formulation and management is a widely dispersed and highly diverse activity. So, ideally, the presence of a template, or model, to guide the process and standardize the day-to-day SCRM activity would be a considerable advantage to any supply chain.

In this chapter, we are going to examine several existing information and communication technology standard models that could be used to structure and oversee practical product supply chain operations. These models are as follows:

- NIST SP 800-161, *Supply Chain Risk Management Practices for Federal Information Systems and Organizations*
- FIPS PUB 200, *Minimum Security Requirements for Federal Information and Information Systems*
- NIST SP 800-53 Rev. 4, *Security and Privacy Controls for Federal Information Systems and Organizations*

Since there are no commonly accepted standards for ensuring supply chain security within the general business world, we will use the established federal guidelines. The practices that are embedded in these standards are only required for compliance with governmental regulations. Nevertheless, any organization that utilizes the principles and recommendations discussed here will find that it is possible to better and more efficiently channel their resources into assuring the security of supply chain operations at a level appropriate for the criticality of its associated products and services.

Federal Information Processing Standard (FIPS) 200 and NIST 800-53 are perhaps the better known and more commonly implemented of these instruments. That is because they have been mandated to ensure comprehensive security across all federal systems. The promulgation of FIPS 200 began in 2006, and the actual beginning of the NIST 800-53 process is traceable to 2011. The practices embedded in FIPS 200 and NIST 800-53(4) specify compliance with the requirements of the Federal Information Security Management Act (FISMA), a federal law that was passed in 2002 as Title III of the E-Government Act.

FIPS 200 and NIST 800-53 Rev. 4 combined specify the control structure and explicit controls to ensure a state of comprehensive information security for all systems utilized by the federal government. To comply with the Act, organizations must first determine the impact level for every one of the security categories listed in FIPS 200. Then the organization utilizes the recommendations specified in NIST Special Publication (SP) 800-53 as a means of selecting and specifying security controls.

NIST 800-161 is a much newer standard, as the numbering suggests; 108 standards were promulgated after 800-53. It provides a set of 21 principles for ICT SCRM. These principles can be utilized to guide the deployment of controls to address the purpose and intent of each principle. The outcome is a comprehensive control array that, unlike the intent of NIST 800-53, provides specifically recommended types of assurance for the communities of practice in a typical ICT product supply chain. The specific tailoring of this standard to the requirements of SCRM makes it so potentially influential in the implementation and conduct of explicit assurance processes for ensuring trusted products.

Why Formal Models Are Useful

In order to effectively oversee their supply chains, organizations need a well-defined and properly coordinated, integrated approach to the identification and management of supply chain risk. The development of organization-wide policy and procedures that outline the roles and responsibilities of all stakeholders is the first step in implementing a SCRM program. Thus, organizations need to first develop a comprehensive ICT product SCRM policy. That policy establishes the organizational structure and defines the necessary roles and responsibilities for implementing systematic supply chain risk identification mitigation activities. Thus, practical operational procedures can be derived, documented, and approved from a consistent and complete framework of governing policies. The procedures should describe who conducts risk assessments, who performs the analysis of the findings, who makes risk decisions, who prepares the formal risk management system, and who specifies any training requirements.

Commercially, the ICT product supply chain encompasses the full gamut of the life cycle and includes the design, development, and acquisition of custom or commercial off-the-shelf (COTS) products. It can also include system integration work and system operation services. It can even comprise product disposal. Any activities of people, the interaction of processes, as well as the services, products, and the elements that make up the product supply chain all have potential impact. Thus, it is necessary to focus on countering supply chain risks throughout the life cycle, not just with respect to the overall acceptance of ICT products and their elements as they are passed down the tiers from supplier to customer. It can also entail managing risks after delivery.

Because there are many tiers and elements in real-world ICT product supply chains, an individual mitigation process or practice will only partly reduce risk to the products that are moving in it. Therefore, a combination of practices has to be integrated into an approach that is appropriately tailored to the context throughout the life cycle of an ICT product. The overall goal is to decrease supply chain risks. Consequently, organizations select the practices based on their suitability for a specific application within the sourced product's performance, cost, and schedule criteria, and the criteria are set by the business and embedded in the contract.

ICT product stakeholders should take the lead in coordinating and implementing the supply chain assurance activities for their particular supply chains. Supply chain assurance is categorized by the degree of reduction of the likelihood and severity of harm if the ICT product supply chain is compromised. Logically, any assessment of that harm should include an assessment of the importance of the system and the impact of the actual compromise on the business goals of the organization's day-to-day operations and assets.

The assessment can also include impacts on individuals and other organizations within the sphere of operation of the product. Thus, organizations need to consider the level of sensitivity of the sourced product's context and operation when developing and applying any SCRM practice. Not every ICT product even needs to implement SCRM strategies. Whether the organization needs to commit resources to the assurance of a given supply chain depends on its impact level. Therefore, prior to making the expensive commitment to manage the operation of a supply chain, the organization should make a clear determination of the level of assurance required.

When information systems services and COTS elements are delivered up a supply chain it is essential that the acquiring organization employ or require standard evidence of good ICT product sourcing practices. Those practices should be well defined and characteristic of commonly accepted good development principles and practices. The implementation of an effective ICT product SCRM process begins with the fundamental performance of those principles and practices. Examples of good practices are as follows:

1. Incorporate the use of multiple suppliers or multiple supply chains.
2. Actively manage suppliers through contracts/service-level agreements (SLAs).
3. Use trusted third-party auditing mechanisms for assessing the product of each step.
4. Perform quality assurance processes for all specified security features.
5. Assign roles and responsibilities and enforce them.
6. Implement an appropriately tailored set of baseline security controls for the process.

Standard models of best practice document and embody the commonly accepted activities outlined in Figure 5.1. Essentially, the adoption of any commonly accepted standard model of best practice becomes a proxy for confirming the correctness of the ICT product supply chain operation because of two factors: First, compliance with a standard process architectural model is often mandated in contracts. More importantly, the standard models discussed in this chapter represent an intentional effort to develop and document an authoritative framework for ensuring comprehensive best practice in ICT product development, acquisition, and sustainment.

NIST SP 800-161, Supply Chain Risk Management Practices for Federal Information Systems

The ICT product supply chain is complex. Therefore, explicit risk management practices are required in order to ensure the appropriate level of assurance for each individual supply chain. One challenge is that extreme diversity in the supply chains, for the impossible number of technological products and services, makes it difficult to utilize the same standard set of practices within the wide range of contexts and communities of practice.

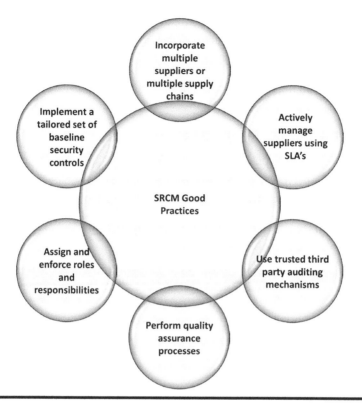

Figure 5.1 SRCM good practice examples.

For example, do you control the COTS elements of the product in the same way as you would unique product elements that are specially developed for a single application? Therefore, the various stakeholders up and down the ICT product supply chain need some common unifying foundation to base their decisions about what practices would best address the risks that are present in their particular ICT product supply chain.

As we have seen, the ICT product supply chain is typically composed of the acquirer, supplier, and integrator communities of practice. Therefore, it is important to be able to detail exactly what protection will be required in each instance and for each appropriate community of practice when implementing the specific practices considered necessary to address risk in a particular supply chain. The requirements for each community of practice need to be captured and documented in the contractual language of the protection scheme for each organization in the ICT product supply chain. This includes specific legal documentation of how the actual controls will be shaped and implemented and where they apply within each community. The guidance for making those decisions is captured in the NIST SP 800-161, *Supply Chain Risk Management Practices.*

Supply chain security is a relatively new concern, and the solutions, methods, techniques, approaches and tools are only beginning to emerge in standard form. Perhaps the most influential of these is the NIST SP 800-161 standard. This particular standard is expected to have wide-scale influence on how controls for SCRM are formulated and implemented through the industry at large.

The 21 Principles for SCRM

The best practices embedded in this standard are designed to promote the trusted acquisition of sourced ICT products from worldwide suppliers, because these practices embody risk management strategies for every element of the total ICT product supply chain operation. In that respect, the NIST SP 800-161 standard embodies 21 fundamental principles for SCRM (Figure 5.2). These are as follows:

	Supply Chain Risk Management Principles of Best Practices
1	Maximize acquirer's visibility into the actions of integrators and suppliers in the process
2	Ensure confidentiality of the uses of individual supply chain components
3	Incorporate conditions for supply chain assurance in requirements specifications
4	Select trustworthy elements and components
5	Enable a diverse supply chain—do not sole source
6	Identify and protect critical processes and elements
7	Use defensive design in component development
8	Protect the contextual supply chain environment
9	Configure supply chain elements to limit access and exposure
10	Formalize service/maintenance agreements
11	Test throughout the system development life cycle (SDLC)
12	Manage all pertinent versions of the configuration
13	Factor personnel considerations into supply chain management
14	Promote awareness, educate, and train personnel on supply chain risk
15	Harden supply chain delivery mechanisms
16	Protect/monitor/audit the operational supply chain system
17	Negotiate and manage requirement changes
18	Manage identified supply chain vulnerabilities
19	Reduce supply chain risks during software updates and patches
20	Respond to supply chain incidents
21	Reduce supply chain risks during disposal

Figure 5.2 SCRM principles of best practices.

1. Maximize acquirer's visibility into the actions of integrators and suppliers in the process.
2. Ensure that the uses of individual supply chain components are kept confidential.
3. Incorporate conditions for supply chain assurance in specifications of requirements.
4. Select trustworthy elements and components.
5. Enable a diverse supply chain—do not sole source.
6. Identify and protect critical processes and elements.
7. Use defensive design in component development.
8. Protect the contextual supply chain environment.
9. Configure supply chain elements to limit access and exposure.
10. Formalize service/maintenance agreements.
11. Test throughout the system development life cycle (SDLC).
12. Manage all pertinent versions of the configuration.
13. Factor personnel considerations into supply chain management.
14. Promote awareness, educate, and train personnel on supply chain risk.
15. Harden supply chain delivery mechanisms.
16. Protect/monitor/audit the operational supply chain system.
17. Negotiate and manage requirement changes.
18. Manage identified supply chain vulnerabilities.
19. Reduce supply chain risks during software updates and patches.
20. Respond to supply chain incidents.
21. Reduce supply chain risks during disposal.

Principle 1: Maximize Acquirer's Visibility into the Actions of Integrators and Suppliers in the Process

It ought to go without saying that acquirers should always seek to obtain maximum visibility into the actions of all the integrators and suppliers and their supporting tiers within any given ICT product supply chain. The obvious purpose is to maintain control over how all product elements are designed, built, verified, validated, delivered, and sustained throughout the element's useful life cycle. That knowledge ensures trust, and to ensure this state of trust, acquirers should always have up-to-date knowledge of the actions of all integrators and suppliers in the ICT product supply chain. They can use the knowledge of integrator and supplier practices to evaluate the trade-offs of performance requirements, the level of risk tolerance or acceptance, and the associated resource constraints.

The primary practices recommended by this principle involve contractual assurance that the integrators and suppliers in the ICT product supply chain provide detailed documentation of the processes they are following to fulfill their contractual agreements. That rule also applies to the integrators and suppliers themselves. The aim is to ensure reasonable visibility and transparency with respect to every element of the product supply chain, including visibility into every supplier they utilize in the development of the product element. Such information may also be important to enable later support should the product require maintenance or change.

Principle 2: Ensure That the Uses of Individual Supply Chain Components Are Kept Confidential

The disclosure of the processes employed and the uses or intentions of a particular element in the ICT product supply chain can lead to unintended consequences. This includes such outcomes as intellectual property theft, business or nation-state espionage, or even loss of business advantage. Therefore, all disclosure of information about practices and uses should be minimized. This is essentially stated in Saltzer and Schroeder's (1974) "least privilege" principle; characteristics and uses of an element should be revealed only to the extent necessary to assure effective achievement of design, development, testing, production, delivery, or sustainment goals for a given element.

This is an information security principle, and the intent is to prevent proprietary information from being combined in such a manner as to compromise the confidentiality of an element or its uses. That includes information about design and development processes, as well as production, test, delivery, or sustainment plans. The standard suggests that the following six items of information should be carefully controlled (NIST 800-161, 2016) (Figure 5.3):

1. What the system is
2. Functions of the system
3. Other systems it will interface with
4. Missions the system supports
5. When or where the system elements will be bought/acquired
6. How many system instances there will be, or where the system may be deployed

The actual rigor of the application of these requirements will vary based on the risk requirements for that particular ICT product supply chain. In the case of national security, these requirements might be enforced by law. However, in the case of conventional, nonsensitive supply chains, it might be possible to exchange information among program elements without strict limits on authorizations.

This final note is an important consideration as acquirers are responsible for understanding and articulating their needs and considering the risks and the benefits associated with releasing or withholding information from their suppliers or integrators. Consequently, withholding information from a supplier or integrator in the ICT product supply chain might make it difficult to create an effective overall comprehensive design for the product. Moreover, a chance for innovation might be missed when a customer withholds pertinent information from the supplier or integrator.

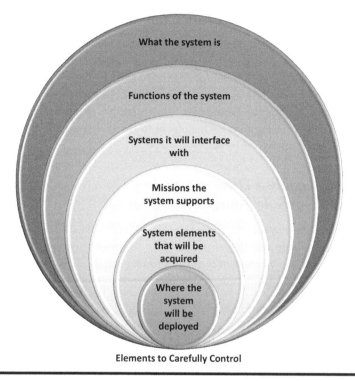

Figure 5.3   Six elements to carefully control.

Principle 3: Incorporate Conditions for Supply Chain Assurance in Specifications of Requirements

There is always an implied obligation to protect the ICT product supply chain against known threats. In the management of the process, this is enabled by thorough and in-depth risk assessment procedures as well as the development of a comprehensive defense-in-depth and defense-in-breadth information security strategy. The formal business goals and objectives, as well as the operational procedures and the attendant technology requirements, must include specific documented ICT product supply chain assurance practices. These in-depth and formally documented requirements will help stakeholders ascertain that the customer organization has thought through and properly expressed to its integrator and supplier communities the requirements for ICT product supply chain assurance. The in-depth articulation of these requirements should ensure that the requirements for supply chain security, as specified in the contract between customer and supplier, have been met and that the chances for unauthorized exposure or access to critical elements or processes in the ICT product supply chain as a whole are suitably considered and contained.

Principle 4: Select Trustworthy Elements and Components

Elements in an ICT product supply chain will inevitably have varying degrees of criticality. This obviously depends on the purpose and utilization of each element. All supply chain elements are potential targets for the injection, intentional or unintentional, of vulnerabilities, and they may be subject to counterfeiting during their life cycle; this must be considered. This principle requires all stakeholders, not just the supplier, to examine each element in the ICT product supply chain to determine its basic trustworthiness.

Principle 5: Enable a Diverse Supply Chain—Do Not Sole Source

An attack is less likely to succeed if the target is dispersed, thereby helping to ensure the availability of required elements and their continued supply in the event of compromise to a given element in the ICT product supply chain. Diversifying the potential sources of supply can make the ICT product supply much more robust. It helps to ensure resistance to attacks by reducing the likelihood or consequences of an attack by enabling alternative sources for the product element. The diversity helps to counter the impact of a failure in a specific element or process by providing back-up alternatives.

Principle 6: Identify and Protect Critical Processes and Elements

This is the defense-in-depth principle; the reason it is important to identify critical processes and elements is to ensure that the most important elements in the ICT product supply chain are protected to the extent allowable with the available resources. It is a key practice in that it applies to all instances of an element within the supply chain. To do this successfully, it is necessary to describe all potential failure modes for each critical element or process within the supply chain, as well as describe how each of these various processes and elements will be affected by any undesirable attack or failure in the ICT product supply chain operation. This requires all relevant stakeholders to develop, refine, and analyze the product tree to the most atomic level of decomposition in order to identify the elements that could cause overall operational failure or compromise security in the prospective supply chain organization.

Principle 7: Use Defensive Design in Component Development

Defensive design minimizes the negative consequences of failure and anticipates the potential ways an ICT product supply chain element or system could fail or be misused. It ensures that the element or system fails securely and makes certain that intentional compromise is difficult or impossible. Defensive design techniques need to be applied by both integrators and suppliers to ensure the integrity of the individual elements and systems that fall within their area of contractual responsibility. Keep in mind that elements may still have intentional or unintentional vulnerabilities even when they originate from trusted suppliers; therefore, defensive design techniques have to be deployed by all suppliers and integrators within a common process architecture in order to ensure against the reduction of risk or damage to every ICT product supply chain element.

Principle 8: Protect the Contextual Supply Chain Environment

Acquirers and integrators should use robust techniques to ensure against unauthorized or unmonitored access to operational processes within the corporate environment; these include ensuring explicit protection for all of the methods that are used for production, assembly, distribution, R&D, test and evaluation, training, and logistics procedures. Competitive advantage in the global business environment dictates that many ICT product supply chain processes and elements are highly distributed. The failure to control physical or logical access to the acquirer, the many potentially diverse integrators, or supplier environments within the ICT product supply chain could result in the injection of malware into systems, elements, or processes; the introduction of counterfeits; or the theft of critical materials and information. Adequate protection should be built into the entire ICT product supply chain process. This is the obligation of all stakeholders in the communities of practice. It is based on the assessment of threat and risk to the ICT product supply chain as a whole.

Principle 9: Configure Supply Chain Elements to Limit Access and Exposure

A product might embody functions that are not needed to fulfill its purpose and is particularly common for COTS elements. The presence of unwanted functionality is difficult to detect because COTS elements are explicitly produced to carry out multiple purposes. One challenge is that unknown and unanticipated functionality can permit unauthorized access or exposure of the system; thus, it is important to configure every one of the elements in the ICT product supply chain to ensure that unknown or unwanted functionality is not implemented in the product that a supplier is developing or supplying.

Principle 10: Formalize Service/Maintenance Agreements

The product sustainment process begins when the element is put into day-to-day use; it generally happens at the end of the acceptance phase of the product delivery process. Once the product is put into operational use, there is often a subsequent requirement for training and long-term support, which is normally specified by contract. The maintenance and operation of the product is termed "sustainment." Rather than being an afterthought, sustainment is an integral part of the ICT supply chain assurance process.

Essentially, the sustainment process must be kept under rational management control throughout its useful life. This includes ICT products that are utilized in conventional business applications for a protracted period. During the life cycle, there is a likelihood that the element will be patched, maintained, and upgraded via the addition of additional enhancements. The long-term process should be kept under strict management control to continuously assure the security of the product after delivery. This is because any change to an element after it has been inspected and approved might provide a new opportunity for subversion of the product through the new parts or services provided up the ICT supply chain.

In this respect, the service and support processes that are utilized throughout the life cycle should also be managed in a way that will limit any new opportunities for unauthorized access or exposure. Thus, sustainment service providers must be considered part of the ICT supply chain.

Principle 11: Test throughout the SDCL

Testing is a critical practice in all forms of technical work. Tests and reviews are used to determine whether a given countermeasure has been properly deployed and works correctly. The purpose of the testing element embedded in any ICT supply chain is to validate compliance with specifications and requirements, establish that the system reacts properly within defined environmental criteria, and find and fix any relevant weaknesses and vulnerabilities in the ICT supply chain element, product, or process.

Testing also shapes the form of the corrective response. It categorizes and analyzes vulnerabilities in the operational deployment of the elements and attempts to understand the potential impacts of attacks and attempts to subvert. Testing applies to all forms of ICT supply chain products including COTS approval, integration processes, and the integration of COTS or custom elements into the evolving product as it moves up the supply chain. Testing applies at many points in the ICT supply chain, from concept to retirement.

Principle 12: Manage All Pertinent Versions of the Configuration

An ICT supply chain entails many components at various levels of integration up and down the process. These comprise a single entity in the final product, but as they progress up the supply chain all components can be represented and interconnected in a multitude of arrangements to meet a variety of business needs. All abstract entities, like software and systems, are in a constant state of evolution and change in response to new components, updates to existing components, new patches, and security threats. To ensure that the precise form of the ICT supply chain product is known and that the state of the operation of all components is well defined, it is necessary to conduct a formal and systematic security configuration management process.

Configuration management is a fundamental discipline in all technical work and is an essential component in an ICT supply chain security risk control process. Configuration management keeps and manages relevant information about product elements throughout the product development process, and this information is maintained as the operational configuration of the product. Given the importance of configuration management, it is important to ensure that the process is conducted securely and correctly in order to ensure ICT supply chain integrity.

Principle 13: Factor Personnel Considerations into Supply Chain Management

ICT supply chains are composed of people, and as such, there is a need for the many diverse people within the supply chain organization to be identified, authorized, and regulated with respect to their work functions. This requires explicit management controls for personnel assurance. Many rules for enforcing management control over the people in an ICT supply chain are specified in the principles of Saltzer and Schroeder as well as in the operational controls of NIST SP 800-53 Rev. 4. This also implies the need for suitably targeted and appropriate security awareness and training.

Principle 14: Promote Awareness, Educate, and Train Personnel on Supply Chain Risk

People are the weakest link in an ICT supply chain, and it is critical for the organization to establish a comprehensive and appropriately tailored awareness, training, and education program. The program should provide sufficient knowledge for all personnel in the organization up-and-down the supply chain to understand all relevant policies and procedures, in addition to managerial, operational, and technical security control practices. These practices should center on ensuring that the members of supply chain organizations have specific awareness, education, and training to fulfill their assigned roles within the ICT supply chain operation.

Principle 15: Harden Supply Chain Delivery Mechanisms

In the end, the overall purpose of an ICT supply chain is to deliver an acceptable product to a customer. Supply chain assurance does not end when the product is completed. Instead, it is necessary to ensure that each element in the overall product tree is addressed in a way that ensures that no security violation or vulnerability is passed up the process from one tier to another.

In essence, the role of this principle is to ensure that malicious or unauthorized individuals are not allowed to gain access to individual product elements as they move up the ICT supply chain. It should be relatively evident that unauthorized access could lead to detrimental modifications of components or outright malicious insertion of harmful objects into the product. Thus, the hardening process has to extend throughout the entire ICT supply chain delivery process, up to the delivery of the system itself.

Principle 16: Protect/Monitor/Audit the Operational Supply Chain System

ICT supply chains are dynamic entities, and they can change, grow, or even degrade based on day-to-day operational activities. Consequently, managers should be able to monitor the operation of the supply chain as it evolves. Therefore, the presence of a set of valid controls and a formally established and well-defined process for monitoring and auditing the operation of all elements at every tier in the supply chain is a necessity. Additionally, some form of audited control-based assurance is necessary because personnel management requires that each organization have the ability to monitor the actions of the individuals in the ICT supply chain.

This type of auditing is not as simple as it sounds. Most big ICT supply chains are multitiered and frequently global. The audit process won't be complete and effective unless it is conducted within the operation. Thus, the audit function needs to be viewed as a complex system of deployed functionality throughout the ICT supply chain, and the audits themselves have to be conducted on a regular and systematic basis.

Principle 17: Negotiate and Manage Requirements Changes

This is known as "requirements management" in many of the old capability maturity models. Requirements management is a necessary control process in the ICT supply chain assurance operation because requirements define the product. Adding additional system or element features typically increases complexity, which in turn creates opportunities for exposure and potential new vulnerabilities. Therefore, any additional requirements or unconstrained changes to requirements can create new or unknown supply chain risks. Thus, each additional requirement that adds elements, element features, interfaces, or new interdependencies or processes, especially those that result in additional suppliers or complexity, must be strictly managed and authorized up and down the ICT supply chain.

Principle 18: Manage Identified Supply Chain Vulnerabilities

Supply chains are complex, and weaknesses and vulnerabilities are the inevitable consequence. Therefore, a process that will securely manage threat identification, analysis, and remediation upon discovery of a supply chain problem is required. That should involve supply chain stakeholders. The stakeholders conduct a formal vulnerability resolution and management process to determine specific actions and allocate resources to address and limit a reported vulnerability's adverse consequences.

ICT supply chain vulnerability management begins when a vulnerability in a supply chain element or component is discovered and identified. Then a determination is made as to the potential impact of exploitation and dependencies across the supply chain. This should include a determination of whether the vulnerability is currently exploited and where the vulnerability was introduced. From the analysis, the level of risk generated by this vulnerability is determined and a decision is made to mitigate or accept the risk.

Principle 19: Reduce Supply Chain Risks during Software Updates and Patches

Software updates and patches are inevitable in any technological product, and a significant challenge is that a patch changes the system. Although patches and changes are utilized to address known problems, changes can introduce new vulnerabilities. A failure to promptly update, or patch, the product ensures that an exploitable vulnerability will remain within the ICT supply chain or its products. Supply chain risks need to be kept in mind and managed when updates and patches are installed in the product.

Principle 20: Respond to Supply Chain Incidents

Incidents happen all the time in technology work. Whether accidental or the actions of an adversary, these incidents can be harmful and in some cases catastrophic. Therefore, a formal and well-defined ICT supply chain incident management process is a necessary part of good SCRM. To ensure the effectiveness of the incident response process, organizations need to plan and implement a process for receiving notice of and subsequently managing supply chain incidents.

Procedures for detecting, reporting, and responding to security incidents must be included in the plan to ensure continuity of operations. The ultimate objective is to conduct the daily operations of the organization's information systems and achieve a level of security that prevents the unauthorized access, use, disclosure, disruption, modification, or destruction of information. The process should include formal mechanisms to detect, report, and analyze incidents and deploy an appropriate response to all incidents occurring within the ICT supply chain.

Principle 21: Reduce Supply Chain Risks during Disposal

Eventually, ICT supply chain products must be properly disposed of. Retirement and disposal is an inevitable consequence of progress; however, the absence of a planned and secure disposal procedure can lead to unauthorized access to components, elements, and products and the disclosure of vulnerabilities and weaknesses that would otherwise be unknown to a potential adversary. It can also lead to the subversion of replacement of subsequent systems or direct attacks on existing ICT supply chains.

Making Control Structures Concrete: FIPS 200 and NIST 800-53(Rev 4)

The standards that dictate how a federal government organization will comply with the FISMA can provide an ideal basis for building on the principles of the NIST SP 800-161 standard. FIPS 200 and its attendant set of NIST SP 800-53 Rev. 4 controls were promulgated after long study as a recommended means to deploy a complete set of assurance behaviors for any ICT operation. The general focus makes them excellent templates for the type of comprehensive process definition framework to guide any ICT assurance process.

NIST SP 800-161 was developed with the detailed control sets of NIS 800-53 Rev. 4 in mind. NIST SP 800-53 Rev. 4 itemizes the assurance control structure for any organization operating within an ICT product supply chain. These controls can be used as a concrete basis for building the management control necessary to formulate practical, real-world SCRM processes.

The FIPS 200 and NIST SP 800-53 models are often utilized to certify legally mandated compliance laws and regulations such as the Sarbanes–Oxley Act (SOX), which applies to the field of financial reporting; the Health Information Portability and Accountability Act (HIPAA), which applies to health care; and FISMA, which applies to all agencies of the U.S. Federal Government. However, the process and the embedded controls can easily apply to creating control assurance frameworks for organizations that operate in any form of ICT supply chain.

FISMA, known officially as Title III of P.L. 107-347, is the enabling legislation that authorized the development of the FIPS 200 compliance model as a proxy for audited assurance of robust security for federal systems. Unlike SOX and HIPAA, which are tailored to specific environments, FISMA is comprehensive legislation that dictates every aspect of correct security practice for a large-scale information system environment. This next section will focus on the best practices recommended for FISMA. The reader needs to understand that the principles and practices outlined in this regulation apply generally to the proof of embodiment of best practice in the ICT product supply chain universe.

FISMA is an element of the E-Government Act, formally known as Title III-Section 301 Information Security. This Act was signed into law in 2002. FISMA formally recognizes the

> "importance of information security to the economic and national security interests of the United States." Consequently, FISMA requires each federal agency to "develop, document, and implement an enterprise-wide program to secure information and the information systems that support the operations and assets of every federal agency, including those provided or managed by another agency contractor, or other source."
>
> *(FISMA, 2002)*

On the surface, FISMA appears to apply only to general assurance of federal information systems, but the preceding quote highlights the importance of the Act to efforts to achieve comprehensive ICT product supply chain assurance. The principle assurance activities required in FISMA generally apply to ICT work and justify the application of the mandated assurance best practices in many other applications, for instance, the assurance and control of private-sector supply chains. The need to practice systematic security within a framework of areas of risk applies to supply chains as much as it does federal systems.

Therefore, the practices that document compliance with FISMA are also excellent proxies for best practice in the general operation of a product sourcing operation. FISMA is implemented by two federal standards, and these standards are issued by NIST.

Application of FIPS 200 and NIST 800-53(Rev 4) to Control Formulation

FIPS 200 defines the general security areas that must be specifically addressed to document systematic assurance within a federal system. In actuality, two different standards are part of the FISMA compliance puzzle. FIPS 199 characterizes the sensitivity of a given system. The results of the FIPS 199 characterization are then formalized as a requisite level of classification: low, medium, or high. The reader should note that we are not discussing the sensitivity classifications in the FIPS 199 standard here, because we are focusing on the generic best practice controls captured in FIPS 200.

FIPS 200 dictates best practice that must be documented as correct for the system to be approved as fit to operate. NIST 800-53 controls for each of the FIPS 200 security areas are dictated by the analogous sensitivity of the system. All categories of FIPS 200 should be addressed by an appropriate control that has been derived from NIST 800-53. FIPS 200 specifies minimum security requirements in 17 security-related domains. Federal agencies must meet these requirements by using the security controls specified in NIST SP 800-53; NIST 800-53 itemizes the current commonly accepted state of proper practice in the safeguards and countermeasures for assurance of an ICT system.

The selection of the security controls for the 17 FIPS 200 areas is dictated by a risk-based approach. FIPS 200 establishes minimum levels of due diligence for assurance for a given system and the ideal control sets specified in NIST 800-53(Rev 4) then serve as a more consistent, comparable, and repeatable basis for specifying the specific assurance controls for the target system. The 17 areas covered by FIPS 200 represent a broad-based response that addresses all aspects of management, operations, and technology. This is important in an ICT SCRM because supply chains are composed of a diverse set of operations. Therefore, a comprehensive set of standard policies and procedures plays an important role in the effective implementation of enterprise-level assurance of the ICT product supply chain. Therefore, the following 17 standard security-related areas that are specified in FIPS 200 are likely to be relevant to some aspects of SCRM:

1. *Access control*—Limit access to authorized users, processes, and devices.
2. *Awareness and training*—Ensure that personnel are adequately trained.
3. *Audit and accountability*—Ensure that actions can be traced to enforce accountability.
4. *Certification, accreditation, and security assessments*—Assure continued effectiveness.
5. *Configuration management*—Establish and enforce baseline configurations for systems.
6. *Contingency planning*—Establish, implement, and maintain plans to ensure continuity.
7. *Identification and authentication*—Authenticate the identities before allowing access.
8. *Incident response*—Establish operations for incident handling and incident reporting.
9. *Maintenance*—Establish controls to perform periodic maintenance.
10. *Media protection*—Limit access to authorized users to sanitize media before disposal.
11. *Physical and environmental protection*—Limit physical access to equipment.
12. *Planning*—Develop plans that describe planned security controls.
13. *Personnel security*—Ensure that personnel are trustworthy, including third parties.
14. *Risk assessment*—Periodically assess the risk to operations.
15. *Systems and services acquisition*—Use adequate measures to ensure sourced products.

16. *Communications security*—Assure transmissions at the perimeter and within boundaries.
17. *System and information integrity*—Identify, report, and correct flaws in integrity.

After categorizing security for its system, the organization selects a set of security controls from NIST SP 800-53 that satisfy the requirements for the 17 categories in FIPS 200, shown in Figure 5.4. The appropriate controls for a given level of sensitivity are specified in NIST SP 800-53. Unless exceptions are granted, organizations must employ all security controls specified for their respective baselines.

The selection of controls for any given area of security should involve the relevant set of stakeholders up and down the supply chain. To ensure a cost-effective, risk-based approach to security across the supply chain, the explicit behaviors that are meant to characterize the desired assurance activities need to be tailored into a comprehensive and coherent security control baseline. That baseline then must be coordinated and authored by the proper decision-making authorities up and down that specific ICT supply chain. The resulting set of operational security controls is then documented, and the requisite behaviors are installed and enforced as standard operating procedure.

FIPS 200 Security Control Categories	
1	*Access control*—limit access to authorized users, processes, and devices
2	*Awareness and training*—ensure that personnel are adequately trained
3	*Audit and accountability*—ensure that actions can be traced to enforce accountability
4	*Certification, accreditation, and security assessments*—assures continued effectiveness
5	*Configuration management*—establish and enforce baseline configurations for systems
6	*Contingency planning*—establish, implement, and maintain plans to ensure continuity
7	*Identification and authentication*—authenticater identities before allowing access
8	*Incident response*—establish operations for incident handling and incident reporting
9	*Maintenance*—establish controls to perform periodic maintenance
10	*Media protection*—limit access to authorized users sanitize media before disposal
11	*Physical and environmental protection*—limit physical access to equipment
12	*Planning*—develop plans that describe planned security controls
13	*Personnel security*—ensure that personnel are trustworthy, including third parties
14	*Risk assessment*—periodically assess the risk to operations
15	*Systems and services acquisition*—use adequate measures to ensure sourced products
16	*Communications security*—assure transmissions at the perimeter and within boundaries
17	*System and information integrity*—identify, report, and correct flaws in integrity

Figure 5.4   FIPS 200 security control categories.

The Generic Security Control Set

The NIST SP 800-53 minimum assurance requirements should specify the security behaviors and activities that the ICT supply chain management has to deploy in order to achieve the minimum degree of compliance with the assurance levels required in a given sourcing contract. The organizations in the supply chain implement these controls and monitor them to provide the level of continuous confidence that the requisite controls were implemented correctly, operate as designed, and produce the desired security outcomes.

The NIST SP 800-53 control activities are specified one control at a time. However, in order to keep the necessary knowledge of the complete set of an ICT assurance measures at a consistent level of awareness, the individual controls are grouped into a single security control baseline. The baseline is homogeneous in that each control within the baseline serves a purpose relating to the larger set of logically designed and interrelated security requirements.

The true advantage of utilizing NIST SP 800-53 to establish a security system rests on the fact that the control set is comprehensive and homogeneous. It has been designed by NIST over a long process involving many experts to achieve the general goal of complete system assurance. Since the standard control baselines within NIST SP 800-53 are complete and correct, their security recommendations provide a standard template for the practical implementation of assurance control in any organization within a supply chain. This means that an organization that adopts a NIST SP 800-53 control set essentially has had much of the concept and design work done for it, even before starting the process of tailoring a practical set of controls for application up and down the supply chain.

NIST 800-53 Control Baselines

The control baselines in NIST SP 800-53 provide a well-defined set of control activities to satisfy the assurance requirements for a given security area in the FIPS 200 standard. Selecting and applying a specific control from 800-53 involves implementing real-world actions that satisfy the intention of that control statement, within that security category. Security categories represent the total set of security requirements for a product at an assessed level of impact. Organizations then tailor the recommended action to their contextual needs to produce a standard baseline of controls that satisfies the generic assurance requirements of the supply chain in which the organization resides. The 800-53 recommended baseline behaviors for an organization describe the safeguards and countermeasures that have been employed to mitigate risks within that organization and by implication all organizations in the supply chain. These baselines are documented and integrated into the overall assurance plan for that ICT product supply chain.

Detail of Controls

In all cases, the organizations in an ICT product supply chain must be able to demonstrate that all of the requisite security controls are in place and generally satisfy the expressed requirements for assurance for that organization's position in the supply chain. The organization must document that each control was developed in a way that ensures a high degree of confidence that the control is complete, consistent, and correct. The organization should also capture and document objective evidence that the control is operating effectively and meets its stated functional requirements.

The desired condition is that the security controls are unambiguously defined and that no obvious errors in definition or execution are known to exist. If a control is shown to be inadequate, then timely remediation is required. To support this latter requirement, the organization provides a description of the control's functional properties as well as its design and development requirements. This description contains sufficient detail to permit subsequent analysis and testing. Organizations must be able to demonstrate that these security controls are present and documented. To assure sufficient reliability of each control, the organization is expected to use formal and well-defined processes to design, develop, and implement its controls. The organization must also produce documentation to support audited proof of compliance with the security requirements

To ensure a basic level of security, an organization must test and assure the correct operation of each control in the moderate baseline through formally planned and executed processes. In general, the organization develops a precise description of all requisite behaviors, technical actions, and activities to ensure that the control will satisfy all intended outcomes when implemented properly. To ensure that the testing and assurance process is performed properly, the organization must develop a specific description of behaviors that the control must exhibit. This description must provide sufficient detail to allow the organization to confirm that the control is functioning correctly.

Thus, the organization must provide a description of expected outcomes for each control's operation. These descriptions and requirements must be objective and detailed enough to allow each control's performance to be quantitatively analyzed and assured. Then, the everyday functions of the security control must produce quantitative data to confirm that its operation complies with stated functional requirements. The organization must also include a description of how it will continually assess and improve the effectiveness of the control. The description must include actions to ensure that the organization develops and records relevant documentation or proof of performance and then performs and records assurance activities to achieve its improvement goals.

Because any set of controls is going to be only as effective as the people who operate them, the organization needs to ensure the performance of the people involved with designing and operating the security controls. This requirement includes providing a detailed policy description of specific staff responsibilities and behaviors. If the documentation and human resource management portion of security system deployment is correctly defined, the organization has an objective basis to ensure that the necessary abilities to operate the security system are always present (NIST SP 800-53 Rev. 4).

Six Feasibility Considerations for NIST 800-53

In order to embed any given set of NIST 800-53 controls within an ICT product supply chain, it is necessary for each of the component organizations to account for specific contextual factors that might influence the deployment of each recommended control. Six large, contextual considerations should be factored into planning to decide whether an issue will influence how the baseline controls are applied. These considerations are (NIST, 2014) (Figure 5.5):

■ *Technological feasibility*—If a security control requires the use of specific technologies and versions, commonsense dictates that the control will work only if they are supported within the existing system. Thus, an organization cannot recommend a control without considering whether the needed technology is in place. Otherwise, for example, the installation of a new control might not work or might cause a major vulnerability in the system due to compatibility problems with existing components.

■ *Compatibility of management processes*—The organization's infrastructure is a system, so a change to one of its constituent processes or activities can cause the same concerns as a change to a technical component. Consequently, an organization must be able to say with assurance that a change to operating processes will not harm security, either by creating new vulnerabilities or by causing a misalignment of processes within the protection scheme.

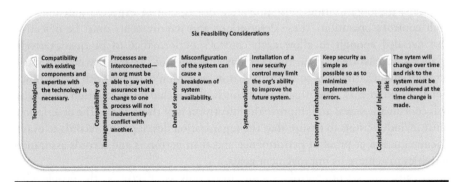

Figure 5.5   Six feasibility considerations.

- *Denial of service*—This consideration is an extension of the previous one. Misconfiguration of the security scheme due to the addition of a new control can lead to a denial of service. A breakdown in system availability is separate from a problem with a process capability because the denial of service can involve either a technical or procedural consideration. If a conflict results from adding an incorrect technical control to the system's protection scheme, the system might stop working or a critical piece of functionality might be lost. On the management side, the addition of a behavioral control that causes a conflict with an everyday business process can cause a security exposure or harm business operations.
- *System evolution*—The installation of a new security control might limit the organization's ability to improve the system for the future. This is a strategic concern because the security of the current system will probably not be affected. Nevertheless, an incorrect decision in the evolution of the system—for instance, selecting a platform with built-in technical limitations—can lead the organization into a cul-de-sac from which it is difficult to recover. Because this feasibility concern involves a certain amount of "fortune telling," it rarely surfaces in security planning decisions. However, the ability to extend a system over the long term is a legitimate issue that needs to be considered when implementing a control.
- *Economy of mechanism*—This concern is perhaps the most pragmatic of all. It enforces an idea that Saltzer and Schroeder call "economy of mechanism," which refers to the commonsense principle of keeping security as simple as possible. If the controls are simple and easy to operate, then the possibility of implementation errors is greatly reduced. Also, any errors that do exist are easier to find and fix. More importantly, the user is less likely to make a mistake in executing a control if it is intuitively obvious to operate.
- *Consideration of injected risk*—Because the system will change and evolve over time, all of the prior five issues could create new or unanticipated risks. The organization must consider the effects of these changes and new risks. A new risk can be mitigated if it is properly considered at the time the change is made, but the timing of the response must be feasible.

Feasibility is an important consideration in security system implementation because there are so many factors involved in the security process, each of which must be satisfied in order to have effective real-world security practice. Therefore, all of the above standard considerations, as well as any unique feasibility concerns that might apply to a specific organization, have to be addressed by the designers of the system. Done properly, the design should always be fully implementable and able to be operated without any interruption to the security the system provides.

NIST 800-53 Catalog of Baseline Controls

NIST SP 800-53 is the basis for selecting and specifying the security controls that are to be implemented in a given organization. As previously mentioned, the standard is specifically designed to support agencies of the federal government; however, the recommendations can be extended to apply to any information system in any environment. The official aim of NIST SP 800-53 is to facilitate a more consistent, comparable, and repeatable approach for selecting and specifying security controls in information systems and to provide a catalog of those controls (NIST SP 800-53 Rev. 4).

The objective of the control catalog is to provide a complete set of prototype controls to enable a comprehensive security response. Those controls are tailored directly from the generic set of controls specified in NIST SP 800-53. The controls must be tailored to satisfy the security requirements of the specific system within the organizational context where they will be applied. The organization selects the appropriate controls and then customizes and implements them to address specific threats.

The NIST SP 800-53 baseline ensures that security controls are defined more consistently across the organization; however, it is necessary to tailor the controls specified in that standard controls baseline to explicitly address all relevant threats and vulnerabilities. The preparation of a tailored response from a standard control set involves execution of a risk analysis process. That risk analysis identifies, characterizes, and evaluates the specific risks a new threat might pose. The risk assessment itself can be performed in a variety of ways depending on the organization's needs. But any process associated with implementing 800-53 should be driven by a comprehensive risk assessment.

The set of security controls recommended by NIST SP 800-53 comprises the current state of the practice for IT risk management. The selection and tailoring of these controls is part of the process of managing organizational risk. The process entails several key phases that include obvious steps such as threat assessments, risk analyses, and the selection and implementation of a corresponding set of controls based on those evaluations. The standard recommends the following set of seven steps for building an effective risk management system:

1. Understand the impact of risk on each system in the organization.
2. Select and baseline a satisfactory set of security controls to address estimated impacts on each system.
3. Adjust or tailor the initial baseline of security controls after assessing the impacts of identified risk on the system's operating environment.
4. Document the security controls in the system security plan, including the organization's justification for any refinements or adjustments to the initial set of controls.
5. Implement the security controls in the system. For legacy systems, some or all of the selected security controls may already be in place.

6. Assess the performance of the security controls to determine that they were implemented and operate correctly and satisfy all documented security requirements.
7. Monitor and assess the selected controls continually.

An organization needs to maintain continual alignment between the control set and the security needs of the asset being protected. Security risks must be categorized because it is essential to align the implemented security measures with the importance of the information they are designed to protect. In other words, the security controls should be proportionate to the potential harm to information that might come from a violation or incident in the controlled area. This ongoing balancing act ensures proper security and maximizes the cost benefit of implementing the controls.

After the organization selects an appropriate security control baseline, it must consult the standard to apply scoping to the initial baseline. Scoping ensures a proper balance between the degree of protection and the assumed level of threat by using the appropriate set of controls based on the assumed sensitivity of the content of the ICT product supply chain. Once the organization has scoped the baseline, it can tailor the actual implementation of controls according to its needs.

After scoping and tailoring, the last step is to specify the practical form of the security controls within the execution of the supply chain operation. The organization uses a risk-driven approach to tailor the control set to the needs of the operating environment. The aim is to ensure that the specification of controls is feasible in terms of technology and available resources.

Implementing Management Control Using the Standard NIST SP 800-53 Rev. 4 Control Set

Management and controls go hand in hand in that a substantive state of management is implemented by the specific control behaviors that dictate the right operational practice. Because the operation itself can be incredibly complex, controls are not implemented in one single monolithic batch. Instead, they are implemented top down in layers. In simple terms, the implementation process is hierarchical, in that the highest level, and most general control statements, which could conceivably be the organization's top-level policies for SCRM, are further elaborated by breaking them down into several smaller, easier-to-implement level controls. The lowest of these control levels ensures the desired outcome of the high-level control statement. This decomposition process creates a classic one-to-many hierarchy that, at its lowest level of documentation, can comprise a very large number of specific behaviors, at the individual organizational level. Nevertheless, it is possible to establish rigorous control over the security practices of each organization in the ICT product supply chain using this logical decomposition process.

Practical Security Control Architectures

The security controls specified in NIST SP 800-53 must be shaped into a well-defined architecture made up of security practices for each organizational situation. The security controls in NIST SP 800-53 have been organized into three categories and 17 classes, or families, for ease of understanding. The three general categories of controls are managerial, operational, and technical. Each of the 17 classes implies a related set of required security behaviors. According to the standard, each control must be documented. This documentation facilitates the decomposition process in that an organization schedules the actual, real-world deployment of the specific class of controls within families. However, many of the security controls for a given control requirement can logically be associated with more than one class.

The practical documentation of the organization's security control architecture involves three key components: a control description, supplemental guidance for application of the control, and a control enhancements section (NIST 800-53, 2014). The control section provides a concise statement of the security capability provided for a component or aspect of the ICT product supply chain. The control section describes security-related activities that need to be carried out to satisfy the intent of the control (NIST 800-53, 2016). A supplemental guidance section provides additional information for a specific security control. Organizations should consider using the supplemental guidance when defining, developing, and implementing novel security controls for a given situation (NIST SP 800-53, 2014).

The control enhancements section describes any security capability that is required to create added functionality for a basic control. Control enhancements also describe additional steps needed to increase the strength of a control. In both cases, the goal is to improve the security of a component within the ICT product supply chain due to the prospect of a known threat.

Control Statements

Control statements are concise declarations of the security capability that must be implemented to protect a particular organization or a larger aspect of the ICT product supply chain. The control statement describes standard, systematic security-related behaviors to be executed by the organization.

The control statements in NIST SP 800-53 describe a range of behaviors for the protection of any type of information system application, including ICT product supply chains. Organizations select controls and tailor operations to align their security controls to meet specific business or operating goals. For example, an organization can specify how often it intends to conduct the threat assessments that will shape the form of the control or how frequently it intends to test the access controls themselves. Once specified, these actions become part of the control's documentation set, and the organization's compliance is directly assessed against the specifications in that documentation. Some options for executing the control might be constrained by a minimum or a maximum number of steps or values (NIST SP 800-53, 2014).

The NIST SP 800-53 control catalog allows a degree of flexibility in tailoring the standard controls. This flexibility lets the organization selectively define how to carry out any set of actions associated with the control. Organizations can tailor the control statement only to specify actions that the control must be able to perform or is not permitted to perform (NIST SP 800-53, 2014).

Supplemental Guidance

Supplemental guidance supplies additional information that might be necessary to clarify the control statement. The standard recommends that "organizations should consider supplemental guidance when defining, developing, and implementing security controls." The aim is to make certain that the control and its purposes are unambiguously understood throughout the organization. This applies both to the organization's implementation of that control and to its day-to-day operational execution.

Control Enhancements

The control enhancements section serves as the "other" category. The standard suggests a complete set of commonly accepted generic controls. Nevertheless, since all ICT product supply chains differ to some degree, it is possible that additional security capabilities will be added to the basic control statement to ensure the effectiveness of the control, as well as to complete the control documentation. Consequently, control enhancements generally specify some form of related control behavior that must be added to the actions of a basic control to increase its scope, or its rigor.

For example, a more robust control might be needed to ensure better protection due to the higher impact of a given threat. A control enhancement might also be needed when the organization seeks additions to a basic control's capabilities based on the outcomes of a risk assessment. Control enhancements are documented for each control in a way that each addition to the basic control functionality can be easily identified during an inspection or audit of the operational control.

Real-World Control Formulation and Implementation

The challenge in formulating and implementing security controls is the basic requirement to identify and implement the right set of controls in a real-world situation. Given the complexity of most organizations, it is a significant challenge to ensure that the most effective controls are in place to address an ICT product supply chain organization's unique requirements. For that reason, the concept of using a standard baseline of "must address" controls as a starting point is very helpful to the implementation process. These standard baselines represent best practice as understood by the experts at NIST. The baselines alleviate a considerable amount of natural inertia when setting up a security control system by documenting a "recommended" initial version (NIST 800-53, 2016).

In this respect, the baseline set of controls comprises the minimum set of security behaviors needed to achieve the requisite level of assurance. For conventional organizations in an ICT product supply chain, these assurance requirements would probably be established based on a comprehensive threat analysis for the supply chain.

Limitations of the 800-53 Approach in SCRM

The determination of the final set of security controls is dictated by the organization's risk environment, which is constantly changing. Therefore, the organization must be agile. The degree of agility required is determined by ongoing assessments of existing security threats. In many cases, additional or enhanced security controls are needed after an assessment to address new threats and vulnerabilities or to satisfy the requirements of a new law, standard, or regulation. An identified threat justifies the addition of a managerial, operating, or technical control.

Because the model 800-53 controls are only meant to specify a general starting point, modifications to the deployed control set will probably be needed to account for all the inevitable exogenous factors within the supply chain. Innumerable threats exist in the global environment of the Internet, and organizations cannot plan for everything. Therefore, they probably will need to employ different or additional security controls from those in the NIST 800-53 baselines. This need is entirely plausible given the diverse nature of organizations and their activities.

Such modifications are almost always driven by a supplemental risk evaluation that leads to tailoring decisions and the eventual specialized controls that must be included in the eventual ICT SCRM plan. The baseline control set provides a standard minimum specification of security controls for the SCRM process. The ICT SCRM design process also deploys a set of specially tailored controls to meet threats that are not part of the standard collection of recommended controls. The only requirement is documentation of how the new control will address the identified risk. Each specially designed and tailored practice is a blend of programmatic activities, validation/verification activities, and assurance requirements, as well as general and technical requirements. The programmatic and validation/verification activities are implemented according to the contract formulated between the acquiring organization and the primary supplier. The term "acquirer" designates the single organizational unit that is acquiring the product or element. The term "element" is used throughout to mean a COTS product and is synonymous with components, devices, products, systems, and materials. An element is part of an information system and may be implemented by products or services.

In many cases, the practice will apply to a software supplier and a hardware supplier. Since most hardware devices contain some level of firmware or software, the document does not differentiate between types of suppliers. The term "integrator" is used to depict a third-party organization that specializes in combining products/elements of several suppliers to produce elements (information systems). The term "supplier" is used to depict an organization that produces elements and provides them to the integrator to be integrated into the overall system; it is synonymous with vendor and manufacturer. Supplier in this document also applies to maintenance/disposal service providers. Appendix A provides a glossary of terms used throughout the document.

A description of its intent is provided at the beginning of each practice. The practice then expands into specific activities and requirements that will aid in obtaining varying levels of supply chain assurance. In some cases, the practice or a portion of the practice may be too costly, not applicable, or not feasible to implement for a FIPS 199 high-impact information system. The practices are recommendations and should be considered on a risk management basis. The practices selected for this document take into account that the organization has a developed and implemented robust information security program and uses NIST guidelines and standards.

Security controls are the specific management, operating, and technical behaviors designed to protect information security in an organization. The generic security controls defined in NIST SP 800-53 are planned and installed using a formal managerial process that produces an explicit architecture for the ICT function for a specific organization. The process is implemented by a plan that ensures all designated security controls are implemented properly and are effective in daily operation. Because the controls must always be effective, the implementation of NIST SP 800-53 is built around periodic assessments of risk and feedback obtained during scheduled preventive maintenance inspections of the effectiveness of each control. The outcomes from those assessments include such estimates as the magnitude of harm that could result from the "unauthorized access, use, disclosure, disruption, modification, or destruction of information and information systems that support the operations and assets of the organization" (NIST 800-53).

Within the larger, strategic management plan for the organization, specifically targeted plans are documented to ensure sufficient security for individual networks, facilities, or information systems (NIST 800-53). The plan also includes periodic testing and reviews to evaluate the effectiveness of all security policies, procedures, practices, and security controls. The frequency of these tests and reviews depends on the risk environment, but they are generally performed at least once a year.

Because a recommendation for remedial action might result from these reviews and tests, a process for planning, implementing, evaluating, and documenting such actions must be included in the organization's overall set of practices.

Chapter Summary

This chapter discusses the mechanisms for establishing a SCRM capability within any organization. The goal in each of these cases is to formulate a practical way for suppliers and acquisition and security professionals to implement a unified process that will allow them to make informed decisions about the trustworthiness of sourced products. We examined several existing standard models that could be used to structure and oversee practical ICT product supply chain operations. These models are as follows:

- NIST SP 800-161, *Supply Chain Risk Management Practices for Federal Information Systems and Organizations*
- FIPS PUB 200, *Minimum Security Requirements for Federal Information and Information Systems*
- NIST SP 800-53 (v4), *Security and Privacy Controls for Federal Information Systems and Organizations*

NIST 800-161 is a much newer standard, as the numbering suggests; 108 standards were promulgated after 800-53. It provides a set of 21 principles for ICT SCRM. These principles can be utilized to guide the deployment of controls to address the purpose and intent of each principle.

FIPS 200 and NST 800-53(4) combined specify the control structure and explicit controls to ensure a state of comprehensive information security for all systems utilized by the federal government. To comply with the act, organizations must first determine the impact level for each of the security categories listed in FIPS 200. Then the organization utilizes the recommendations specified in NIST SP 800-53 as a means of selecting and specifying security controls.

The ICT product supply chain is complex. Therefore, explicit risk management practices are required to ensure the appropriate level of assurance for each individual supply chain. The best practices embedded in this standard are designed to promote the trusted acquisition of sourced ICT products from worldwide suppliers. The NIST 800-161 standard embodies 21 fundamental principles for SCRM. These are as follows:

1. Maximize acquirer's visibility into the actions of integrators and suppliers in the process.
2. Ensure that the uses of individual supply chain components are kept confidential.

3. Incorporate conditions for supply chain assurance in specifications of requirements.
4. Select trustworthy elements and components.
5. Enable a diverse supply chain. Do not sole source.
6. Identify and protect critical processes and elements.
7. Use defensive design in component development.
8. Protect the contextual supply chain environment.
9. Configure supply chain elements to limit access and exposure.
10. Formalize service/maintenance agreements.
11. Test throughout the system development life cycle (SDLC).
12. Manage all pertinent versions of the configuration.
13. Factor personnel considerations into supply chain management.
14. Promote awareness, educate, and train personnel on supply chain risk.
15. Harden supply chain delivery mechanisms.
16. Protect/monitor/audit the operational supply chain system.
17. Negotiate and manage requirement changes.
18. Manage identified supply chain vulnerabilities.
19. Reduce supply chain risks during software updates and patches.
20. Respond to supply chain incidents.
21. Reduce supply chain risks during disposal.

The standards that dictate how a federal government organization will comply with FISMA (2002) can provide an ideal basis for building on the principles of the NIST 800-161 standard. FIPS 200 and its attendant set of NIST 800-53 controls were promulgated after long study as a recommended means to deploy a complete set of assurance behaviors for any ICT operation. That general focus makes them excellent templates for the type of comprehensive process definition frameworks for guiding any ICT assurance process.

In fact, NIST 800-161 was developed with the detailed control sets of NIS 800-53(4) in mind. NIST 800-53 itemizes the requisite assurance control structure for any organization operating within an ICT product supply chain. These controls can be used as a concrete basis for building the management control necessary to formulate practical, real-world SCRM processes.

The FIPS 200 and NIST 800-53 models are often utilized to certify legally mandated compliance laws and regulations such as SOX, HIPAA, and FISMA. However, the process and the embedded controls can easily apply to creating control assurance frameworks for organizations that operate in any form of ICT supply chain.

FIPS 200 defines the general security areas that must be specifically addressed to document systematic assurance within a federal system and dictates best practice that must be documented as correct for the system to be approved as fit to operate (FIPS, 2016). NIST 800-53 controls for each of the FIPS 200 security areas are dictated by the analogous sensitivity of the system. All categories of FIPS 200 should be addressed by an appropriate control that has been derived from NIST 800-53. FIPS 200 specifies minimum security requirements in 17 security-related domains. Federal agencies must meet these requirements by using the security controls specified in NIST SP 800-53; NIST 800-53 itemizes the current commonly accepted state of proper practice in the safeguards and countermeasures for assurance of an ICT system.

The selection of controls for any given area of security should involve the relevant set of stakeholders up and down the supply chain. To ensure a cost-effective, risk-based approach to security across the supply chain, the explicit behaviors that are meant to characterize the desired assurance activities need to be tailored into a comprehensive and coherent security control baseline. That baseline then must be coordinated and authored by the proper decision-making authorities up and down that specific ICT supply chain. The resulting set of operational security controls is then documented, and the requisite behaviors are installed and enforced as standard operating procedure.

The NIST SP 800-53 minimum assurance requirements should specify the security behaviors and activities that the ICT supply chain management should deploy to achieve the minimum degree of compliance with the assurance levels required in each sourcing contract. The organizations in the supply chain implement these controls and monitor them to provide the level of continuous confidence that the requisite controls were implemented correctly, operate as designed, and produce the desired security outcomes.

The NIST SP 800-53 control activities are specified one control at a time. However, to keep the necessary knowledge of the complete set of ICT assurance measures at a consistent level of awareness, the individual controls are grouped into a single security control baseline. The baseline is homogeneous in that each control within the baseline serves a purpose relating to the larger set of logically designed and interrelated security requirements.

The true advantage of utilizing NIST SP 800-53 to establish a security system rests on the fact that the control set is comprehensive and homogeneous. It has been designed by NIST over a long process involving many experts, to achieve the general goal of complete system assurance. Since the standard control baselines within NIST SP 800-53 are complete and correct, their security recommendations provide a standard template for the practical implementation of assurance control in any organization within a supply chain. This means that an organization that adopts a NIST SP 800-53 control set essentially has had much of the concept and design work done for it, even before starting the process of tailoring a practical set of controls for application up and down the supply chain.

NIST SP 800-53 is the basis for selecting and specifying the security controls that are to be implemented in each organization. As mentioned earlier, the standard is specifically designed to support agencies of the federal government. However, its recommendations can be extended to apply to any information system in any environment. The official aim of 800-53 is to facilitate a more consistent, comparable, and repeatable approach for selecting and specifying security controls in information systems and to provide a catalog of those controls (NIST 800-53).

The objective of the control catalog is to provide a complete set of prototype controls to enable a comprehensive security response. Those controls are tailored directly from the generic set of controls specified in NIST 800-53. The controls must be tailored to satisfy the security requirements of the specific system within the organizational context where they will be applied. The organization selects the appropriate controls and then customizes and implements them to address specific threats.

The practical documentation of the organization's security control architecture involves three key components: a control description, supplemental guidance for application of the control, and a control-enhancements section (NIST 800-53, 2016). The control section provides a concise statement of the security capability provided for a component or aspect of the ICT product supply chain. The control section describes security-related activities that need to be carried out to satisfy the intent of the control (NIST 800-53, 2016). A supplemental guidance section provides additional information for a specific security control. Organizations should consider using the supplemental guidance when defining, developing, and implementing novel security controls for a given situation (NIST 800-53, 2016).

The control enhancements section describes any security capability that is required to create added functionality for a basic control. Control enhancements also describe additional steps needed to increase the strength of a control. In both cases, the goal is to improve the security of a component within the ICT product supply chain due to the prospect of a known threat.

The challenge in formulating and implementing security controls is the requirement to identify and implement the right set of controls in a real-world situation. Given the complexity of most organizations, it is a significant challenge to ensure that the most effective controls are in place to address an ICT product supply chain organization's unique requirements. For that reason, the concept of using a standard baseline of "must address" controls as a starting point is very helpful to the implementation process. These standard baselines represent best practice as understood by the experts at NIST. The baselines alleviate a considerable amount of natural inertia when setting up a security control system by documenting a "recommended" initial version (NIST 800-53, 2016).

In this respect, the baseline set of controls comprises the minimum set of security behaviors needed to achieve the requisite level of assurance. For conventional organizations in an ICT product supply chain, these assurance requirements would probably be established based on a comprehensive threat analysis for the supply chain.

Key Concepts

■ SCRM is an essential part of doing business in a global economy.
■ Supply chain control structures should comprise a single controlling entity.

- Supply chains are built through transparency created by effective control structures.
- Explicit SCRM control structures can be built using three commonly accepted standards: NIST 800-161, FIPS 200, and NIST 800-53(Rev 4).
- The FIPS 200 standard provides 17 control areas required for proper risk management.
- The NIST 800-53 standard defines explicit controls to implement each of these 17 areas.
- The recommendations in NIST 800-161 are geared to the control structure created by FIPS 200/NIST 800-53.
- Formal control frameworks ensure a stable process for SCRM.
- The creation and documentation of an explicit control framework is a requirement for proper SCRM.
- The NIST 800-161 standard specifies 21 principles for managing the supply chain communities of practice.
- Supply chains rely on contracts to enforce trust.
- People are the weakest link in a supply chain.

Key Terms

acquirer: the entity in the supply chain process who purchases a given product or service

analysis: an explicit examination to determine the state of a given entity or requirement

assurance: the set of formal processes utilized to ensure confidence in a supply chain product

best practice: the execution of a set of behaviors that are both well defined and commonly accepted as correct in each community of practice

community of practice: A known and commonly accepted collection of organizations all performing a given product development or service activity among which a set of formally defined relationships exists

configuration management: rational control of change based on a formal process

consumer: the final customer in the Acquisition process; the entity for whom the product was built

control framework: a comprehensive set of standard behaviors intended to explicitly define all required processes, activities and tasks for a given field, or application

controls: technical or managerial behaviors that are put in place to ensure a given and predictable outcome for a given project, or business goal

incident response: steps taken to reduce the impact of a given event

infrastructure: a planned entity with a consciously designated purpose

monitoring: specific oversight created by a planned collection and analysis of data

outsourcing: the letting of subcontract to third parties to provide a product or service

process: a collection of practices designed to achieve an explicit purpose

process design: the act of translating a given product development into specific steps along a timeline for an intended and verifiable outcome

risk: a given threat with a known likelihood and impact

SCRM: assurance that all elements in the supply chain are controlled and their present status is known

sourcing: the organizational process of product acquisition either through a development process or by the purchase of a COTS solution

testing: validation of performance of a target object

transparency: full and complete understanding of the execution and outcomes of a process among a defined set of stakeholders or partners

verification and validation: the reviews and tests used to ensure a given set of criteria

References

Federal Information Processing Standard (FIPS) E-Government Act (Public Law 107-347). Federal Information Security Management Act (FISMA), 2002.

Federal Information Processing Standard (FIPS) 200, Minimum security requirements for Federal Information and Information Systems, National Institute of Standards and Technology, 2006.

National Institute of Standards and Technology (NIST), NIST 800-53 Rev. 4, Security and privacy controls for Federal Information Systems and Organizations, 2014.

National Institute of Standards and Technology (NIST), NIST 800-161, Supply chain risk management practices for Federal Information Systems and Organizations, 2016.

Saltzer, J. H., and Schroeder, M. D., The protection of information in computer systems. *Communications of the ACM,* 17(7), 388–402, 1974.

Chapter 6

Control Sustainment and Operational Assurance

At the conclusion of this chapter, the reader will understand the following:

- The role of long-term sustainment in supply chain risk management (SCRM) processes
- The concerns and issues of long-term sustainment
- The importance of configuration management in sustaining supply chain trust
- The importance of stable control baselines in maintaining supply chain integrity
- The five fundamental principles for sustaining SCRM
- The detailed activities of change management
- The role and importance of evolution in sustaining effective supply chains

Sustaining Long-Term Product Trust

The term "entropy" describes the tendency for any system to get disorderly over time. Entropy applies to complex things like the universe or simple things like your kid's bedroom. As with any other principle, the effects of entropy are inevitable especially in the complicated world of information and communication technology (ICT) supply chains. This chapter will discuss the ways that the organization can bring order to the disorderly world of global ICT product supply chains with a process called *sustainment*.

With most developed products, sustainment begins after delivery and acceptance. However, with ICT product supply chains, sustainment starts when the actual sourcing process begins. This is due to the supply chain being a living, dynamic thing subject to any number of unanticipated forces that might cause change and that should be controlled in a rational and systematic way to ensure against exploitations and potential breakdowns. The main purpose of sustainment activities is to assure the letter and spirit of contractual requirements throughout the ICT product sourcing process.

ICT products are acquired to facilitate organizational goals with each product having a fundamental set of security requirements. The security environment is continuously changing; so, if some aspect of the evolving product does not meet security requirements, the organization needs to realign the product and/or its processes to bring it back into proper alignment. Environmental factors that impact security include the following:

■ Changes in the security goals, risk tolerance, risk mitigation, or risk acceptance posture of any participant in the supply chain including the customer
■ Changes in attack patterns and/or adversaries
■ Changes in the business model or business environment of a supply chain participant
■ Changes in the motivation, rigor, or extent of the security functions within a supply chain participant
■ Alterations in the staffing or resource base of a supply chain participant
■ Changes in the basic conditions of the participant environment such as production breakdowns or natural disasters
■ A need to make a basic alteration in the design or assumptions underlying the development process

Some of these factors are shown in Figure 6.1. They and any others need to be accounted for in a real-world ICT product supply chain. The accounting should take place at both the overall supply chain system process level and within the processes of any given member organization within that sourcing chain. Consequently, a well-defined and properly stipulated process for managing change, both short and long terms, should be included in any ICT product sourcing system. This involves a dedicated six-step process to

1. Establish and maintain situational awareness.
2. Analyze reported vulnerability and understand operational impacts.
3. Obtain management authorization to remediate.
4. Manage and oversee the authorized response.
5. Evaluate the correctness and effectiveness of the implemented response.
6. Assure the integration of the response into the larger supply chain process.

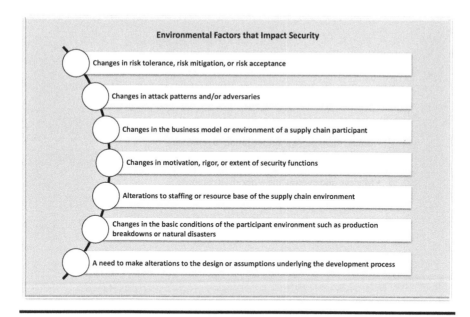

Figure 6.1 Environmental factors that impact security.

The ICT product sourcing and development environment is in constant flux. New threats emerge, and environmental factors change. Consequently, trust in the secure functioning of the ICT product supply chain should be renewed on a regular basis or as needed. Because each organization in the supply chain is closely interconnected to others, both downstream and upstream, an appropriate level of confidence must be maintained up and down the sourcing process (Figure 6.2).

Step 1: Establish and Maintain Situational Awareness

Sustainment creates situational awareness for the supply chain and monitors the hazard space to ensure that there is no present threat to the supply chain's ability to deliver a secure, contracted product or service. Sustainment identifies and records threats or problems, analyzes those problems, and takes the appropriate adaptive, perfective, or preventive corrective action to ensure that the assurance goals of the ICT product supply chain are achieved. The assurance is underwritten by the steps the organization takes to detect and react to all legitimate forms of threat and security violations. The steps are either proactive or reactive. Proactive assurance focuses on three types of actions:

1. Quantitative evaluation of individual supplier products, or their processes, to identify prospective threats or vulnerabilities in those products or processes

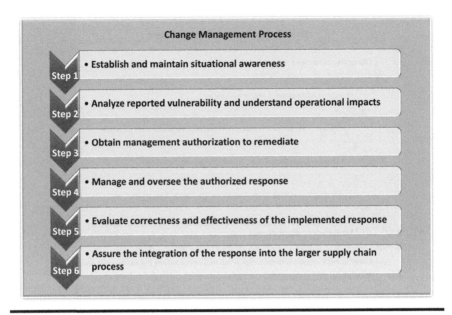

Figure 6.2 Change management process.

2. Developing, evaluating, and perfecting small-or large-scale strategic alterations to a given ICT product supply chain
3. Implementing targeted controls to protect the supply chain and the products moving within it from a known threat

If proactive change is required, then decision-makers have the option to authorize a preventive, perfective, or adaptive response. Preventive responses identify and detect latent vulnerabilities that can be exploited by a known threat and implement a practical solution. Perfective responses are carried out to update and improve the performance, dependability, and maintainability of a given ICT product. Adaptive responses adjust the ICT product to a new or changed environment (Figure 6.3).

If the change undertaken is in reaction to an event that has already occurred, then decision-makers have the option to authorize corrective or emergency action. Corrective responses identify the source of the exploitation and remove the defect or vulnerability that caused it to occur. Emergency changes are exactly as the word implies: unscheduled corrective actions performed on a *do-it-now* basis and often small in scope. The goal of the sustainment process is to ensure unquestioned integrity and trust, up and down the ICT product supply chain.

The first step in the process requires that the stakeholders maintain a continuous an accurate understanding of the threat space. Threats that are known can be countered. Threats that are either unknown or unanticipated cause problems; therefore,

Figure 6.3   Actions of proactive assurance.

sustainment presupposes complete, accurate, and unambiguous knowledge of the current state of the ICT product supply chain at its most basic level of operation. In addition to detailed knowledge of the current state of the supply chain, it is critical to link the details of the overall supply chain operation to the organization's business goals and environment. To get to an acceptable level of detailed understanding, it is necessary to reduce the overall product into its constituent components and supply chain levels.

The eventual delivered product is decomposed top-down from the complete item through its major subcomponents and then down through lower logical stages to the most fundamental level of the supply chain. This breakdown into product components creates a product tree with the individual elements arrayed in a detailed view of the finished product. Then, the individual components are arrayed within that structure, and are audited to characterize their state of compliance with contract requirements. The individual compliance audits provide complete and fully documented assurance of the individual state of each component within the product supply chain tree. In this respect, the drilling down to each element of the tree structure provides the detailed status of the actual condition of the ICT product at each relevant level in the supply chain.

At the same time, the overall product tree structure provides complete representation of the product and its component interrelationships. The essential starting point in any sustainment activity is to obtain and keep a detailed understanding of what threatens the ICT product supply chain. The function that maintains this understanding is a formally structured situational awareness activity that identifies, analyzes, and reports suspicious or threatening occurrences. The goal of situational awareness is to detect and categorize all potential threats, violations, or vulnerabilities that may impact the ICT supply chain performance.

Situational awareness implies the act of continuously monitoring the hazard space as well as testing and assessment of any new discovery or altered condition. If a new hazard or violation is identified, the sustainment function performs an appropriate analysis of the phenomenon to characterize its likelihood of occurrence and impact. Once the analysis is complete, the relevant stakeholders are consulted so they can understand their risk mitigation strategy and criteria and a proposed response is formulated. The response mirrors the strategy as well as the business context that governs whether the risk will be accepted, mitigated, or transferred. The substantive response should achieve the optimum balance between potential harm and the cost of the response. Once the option that the stakeholders have chosen has been authorized and implemented, the sustainment function monitors the change in order to ensure its effectiveness as well as the proper reintegration into the supply chain space.

One of the essential requirements of sustainment is that the stakeholders, not the technical staff or security managers, are always in the driver's seat. This is an essential condition because only the stakeholders understand the nature of their own business processes and goals. This principle is also important because the stakeholder is often the person responsible for staffing and resourcing. If there is buy-in on all sides and at all levels, it will be easier to trust that the change is properly executed and the outcomes are as anticipated.

Due to the constant monitoring of the threat environment, situational awareness is primarily a scanning activity. The visitational awareness function continuously scans the organization's operating environment to ensure that new or changed threats within the information and technology product supply chain are identified early enough to allow an effective response to be prepared. This is a form of operational assurance in the sense that any identified system, data, policy, or user security controls vulnerability must receive a response.

The process itself extends into the organization's operating environment as far as necessary to account for every potential attack vector. Supply chain vulnerabilities can be associated with the technology, the organizational policies and procedures, the actual security mechanisms, the physical security, and the employee actions within any supply chain element. Technical assurance practices include intrusion detection, penetration testing, and violation processing. These are carried out on a routine basis to assess the effectiveness of existing security mechanisms. They are particularly effective where established standards, such as NIST SP 800-53, are utilized as a point of reference (NIST, 2013).

Reviews are also useful and can consist of walkthroughs, inspections, and audits. They can be both managerial and technical. Any threats, vulnerabilities, and violations identified in the testing or review process are recorded and reported using a defined problem-reporting and incident-response process. Because most incidents are not easily defined or routine, the incident response process should be a formally designed and authorized function. Incidents are rarely isolated, and because they can ripple through the supply chain, the incident process itself must be tailored to allow decision-makers to authorize multidimensional actions across a global spectrum.

Step 2: Analyze Reported Vulnerability and Understand Operational Impacts

The impacts of any unforeseen event or change to the ICT product supply chain must be fully understood to manage them wisely. Given this precondition, the first requirement is that the current state of the ICT supply chain must be fully documented and understood. This is the role of the situational awareness in step 1. To ensure the proper, ongoing management of the day-to-day operational process, it is necessary for the organization to take seven standard precautions to analyze and manage change (Figure 6.4):

1. Establish a feasible supply chain, with clear component boundaries, roles, and responsibilities.
2. Document both the general and the unique accountabilities of each component.

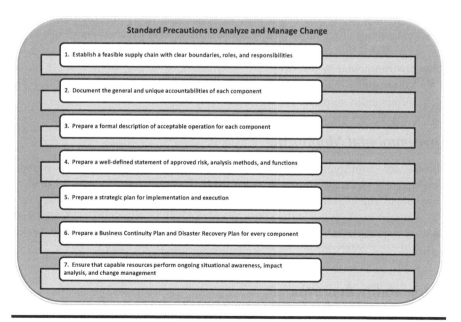

Standard Precautions to Analyze and Manage Change

1. Establish a feasible supply chain with clear boundaries, roles, and responsibilities

2. Document the general and unique accountabilities of each component

3. Prepare a formal description of acceptable operation for each component

4. Prepare a well-defined statement of approved risk, analysis methods, and functions

5. Prepare a strategic plan for implementation and execution

6. Prepare a Business Continuity Plan and Disaster Recovery Plan for every component

7. Ensure that capable resources perform ongoing situational awareness, impact analysis, and change management

Figure 6.4   Standard precautions to analyze and manage change.

3. Prepare a formal description of acceptable operation for each component in the supply chain and the overall supply chain in general.
4. Prepare a well-defined statement describing the approved risk identification and analysis methods and functions.
5. Prepare a strategic plan for implementation and execution of the ICT product supply chain operation.
6. Prepare a business continuity plan with recovery time objectives, network recovery objectives, and recovery point objectives for every component of the ICT product supply chain as well as the process in general.
7. Ensure that adequate resources to capably perform ongoing situational awareness, impact analysis, and change management are provided and that the staff is fully trained.

Threats can arise in the organizational ecosystem at any point in time, and they can represent a range of unanticipated outcomes. Therefore, the conjoined situational awareness and operational change analysis and management functions must be executed on a continuous, disciplined basis.

It is good practice to create a well-defined operational analysis and response management function and make it continuously available to perform impact analyses and direct the requisite change management process to ensure the ongoing stability and security of the ICT supply chain. The aim of this process is to actively analyze and devise countermeasures to address all known, or realistically possible, security threats and vulnerabilities as they are identified as well as to manage the configuration of the ICT supply chain to ensure its ongoing integrity and correctness. The three components of the impact analysis and management function are as follows:

1. Environmental monitoring
2. Vulnerability reporting
3. Vulnerability response management

Environmental Monitoring

Threats are continuously present in every organization's cyber ecosystem. Consequently, the environment in which the ICT supply chain operates is the ideal place to spot and counter imminent attacks. Given this assumption, it is good practice to continuously monitor the ecosystem of the ICT supply chain to identify and respond to prospective threats and attacks as they arise, rather than after they've occurred.

Vulnerability Reporting

The environmental monitoring function is directly linked to the formal process by which any potentially harmful event is reported. This is a standard and disciplined reporting process. The aim is to proactively identify threats arising from defects, malfunctions, and incidents that are identified in the ICT product supply chain. Vulnerability reporting seeks to provide the necessary information to mitigate these vulnerabilities as close to the impending attack as possible. The documentation and reporting process must be well defined in its practices and standard in its execution. The reporting process must be commonly known and accepted within the organization. Therefore, a good practice is to institute a systematic procedure to document and record threat exposures and vulnerabilities.

Vulnerability Response Management

Operators of ICT product supply chains who do not actively seek to manage vulnerabilities identified in the operation of the supply chain and its components, or who do not report vulnerabilities that arise, increase the chances that the final product will be compromised and delayed. On the other hand, the vulnerability response operation must be carefully coordinated and managed to accurately and effectively target complex problems in an often-global context. It is generally considered good practice to establish a central management entity to conduct the identification, reporting, and organizational response to vulnerabilities.

It is a prerequisite of the process that all known vulnerabilities be responded to and addressed. The classification, prioritization, and development of remediation options should be managed, and the best way to ensure a coordinated response is to centralize the effort. Consequently, the actual execution of the vulnerability response process should be overseen by a single authority within the supply chain. This is often a role associated with the primary contractor with the aim to expedite any reported deviations from defined criteria for performance and specified practice.

Ideally, these are reported to the centralized management authority to expedite the remediation. Generally, the responsibility of that authority is to assess and audit the policies, procedures, tools, and standards used for operational assurance of the ICT product supply chain. Additionally, the authority is obligated to document the results of assessments and audits and make recommendations for improvement in the supply chain process.

The overall aim of vulnerability management is to assure the proper performance and ongoing integrity of the entire ICT product supply chain, and the most important responsibility of the authority is to formulate the necessary assurance case. The assurance case comprises the wide variety of evidence that is relevant to certifying the correctness of the ICT product supply chain being overseen. That case guides the assurance process and provides both the purpose and justification for a large-scale activity that might be otherwise considered a costly luxury.

Creating and maintaining an assurance case for a diverse supply chain involves many related activities. The first requirement is to document the evidence of an organizationally standard control compliance process within each of the supply chain components. This includes anticipated outcomes and a means of documenting the achievement of these control objectives. A minimum set of detailed controls should assure every activity within the operation. A well-defined set of organizationally standard actions, procedures, and protocols should specify how to deal with hazards as they are identified (including a description of how to report incidents as they occur or request changes as they are needed and a means for responding to each report or request for change).

Given that two of the five general hazards for supply chains pertain to denials of service, there must also be a formal process for ensuring that the business continuity plan for the entire ICT product supply chain, as well as each of its individual components, is established and kept up to date. The plan is built around the assurance that all relevant participants in the ICT supply chain continuity process know precisely what activities they must carry out in the case of an interruption as well as the timing requirements.

Finally, there are knowledge and awareness components in vulnerability management. Therefore, the overall process should be built around a set of precise steps to ensure the proper level of knowledge and awareness. Additionally, the program often has associated enforcement requirements, which are often satisfied by the implementation of a formal employee education and training program.

Step 3: Obtain Management Authorization to Remediate

In its most fundamental sense, supply chain sustainment is built around the process of change management. Essentially, all changes made in response to the identification of new threats or the need for different practice, must be authorized by the organizationally sanctioned decision authority given responsibility for approval. The approval for change should be based on an intelligent understanding of the costs and benefits. Consequently, step three is fundamentally a policy support function where capable people perform the necessary qualitative and quantitative analyses to support the decision.

Once a report of a vulnerability or need for enhancement is identified, analysts evaluate the impact and the likelihood of potential harm along with the resource commitment required to make the change. Essentially, the projected resource commitment to effect a change is traded off against the cost of exploitation of the identified flaw, and the trade-off analysis is necessary to ensure proper response utilization.

Analysts identify and characterize the affected element of the organization and system. This includes affected policies, processes, their dependencies, and any ripple effects. The identified area of impact or exploitation is then studied from a technical and business perspective to determine the scope of the organizational and system boundaries as well as to identify and describe the precise number of components within those boundaries and any interfacing components and systems. Then a specific change recommendation is developed, which can be either a policy and procedure recommendation or an engineering design process. This formal step is required prior to change approval with the outcome of this action being a well-defined, explicit, and implementable recommendation for change or an enhancement to an existing organizational process or system.

Security and safety impacts of the change must be fully examined and documented. The documentation set, plans, and analyses are communicated to the centralized management entity (step 3: vulnerability response management) for management authorization. Additionally, to determine all potential impacts, the documentation involved in the decision process is formally recorded and maintained in a permanent repository for retrospective analysis. The contents of the repository increase the overall understanding of the control structure of the organizational entity being managed. This wealth of retrospective data supports quantitative and qualitative causal analysis that is generally required to understand the security and control issues associated with the evolving organizational security response.

Understand Impacts

It should go without saying that proper management of the change requires a thorough understanding of all the affected organizational components and objective conclusions about the consequences of the proposed change to each of the affected items. This degree of understanding requires detailed knowledge of all aspects of the policy and process architecture and the affected organizational units and systems. Therefore, it is generally considered best practice to utilize a standard approach and method for the analysis and documentation of both the request and the resulting findings. Because data makes the process objective, all data should be kept in a standard repository.

Once the problem is understood, the analyst seeks to verify the practical conditions that motivated the change request in the first place. This includes replicating the circumstances within the threat ecosystem with the aim to verify the precise nature and status of the issue that requires a response. The verification is necessary to ensure an intelligent and disciplined recommendation for change. Obviously, the organization cannot spend its precisions resources chasing its tail, so the precise situation must be understood to provide a context for the recommendation. Thus, the analyst will seek to verify the nature and threat status of the reported vulnerability. The goal is to fully and completely understand the real-world, everyday security impacts of the reported vulnerability, which is necessary to develop a targeted response strategy.

It should be apparent that the elements of the existing organizational component or system that needs to be modified must first be identified to effect the change. Once all of the processes, procedures, and system components that must be changed are identified and understood, the next step is to develop a specific response strategy. This strategy is applicable to each element, both components and their interfaces. The strategy should embody all of the fundamentals of best practice and should ensure that all relevant security policies, procedures, and system components within the ICT product supply chain be identified, validated, and modified to achieve the desired security and safety outcomes. It is also advisable that the change recommendations take into consideration any relevant legal, regulatory, and forensic requirements. The analysis must ensure proper subsequent change implementation decisions.

To tailor an effective implementation response, it is necessary to know what the implications of a given response strategy or an action taken in response to threat might be. The necessary level of in-depth knowledge implies the need for a comprehensive and detailed impact analysis. Because it will support important decisions about change, the analysis should be based on a well-defined and proven methodology. The aim is to ensure comprehensive and unambiguous understanding of all operational impacts of the change including implications for the associated architecture. The purpose is to make certain that the right change option is selected for recommendation to the central vulnerability management authority. Accordingly, the following 18 decision factors must be considered for each potential remediation option; they are also shown in Figure 6.5:

Decision Factors to Consider for Effective Remediation	
1.	A description of all feasible options for remediation
2.	A detailed project plan for each option
3.	Outcomes of full risk identification, including type and extent of risk for each option
4.	The likelihood and feasibility of each identified risk for each decision option
5.	The exposure and vulnerability type in which the remediation will address
6.	The scope of the proposed remediation with costs and benefits fully itemized
7.	An assessment of the criticality of the exposure or vulnerability
8.	The safety and security impacts if the remediation option is implemented
9.	The safety and security impacts if the remediation option is *not* implemented
10.	A detailed description of the financial impacts for each remediation option
11.	A detailed description of the feasibility and timelines for implementing each option
12.	The impact of the remediation option on the overall security and control architecture
13.	A description of the remediation option's impact on the supply chain policy
14.	The ROI for each option including total cost of ownership and marginal loss percentage
15.	Requisite resource requirements, staff capability, and feasibility of management control
16.	Description of automated security features to be implemented
17.	The impact of the proposed remediation on the assurance case
18.	Description of the impact of a given remediation option on the business continuity strategy

Figure 6.5   Decision factors to consider for effective remediation.

1. A description of all feasible options for remediation
2. A detailed project plan for each option
3. Outcomes of full risk identification, including the type and extent of risk for each option
4. The likelihood and feasibility of each identified risk for each decision option
5. The exposure and vulnerability type the remediation will address
6. The scope of the proposed remediation with costs and benefits fully itemized
7. An assessment of the criticality of the exposure or vulnerability
8. The safety and security impacts if the remediation option is implemented
9. The safety and security impacts if the remediation option is *not* implemented
10. A detailed description of the financial impacts for each remediation option
11. A detailed description of the feasibility and timelines for implementing each option
12. The impact of the remediation option on the overall security and control architecture
13. A description of the remediation option's impact on the supply chain policy
14. The return on investment (ROI) for each option including total cost of ownership and marginal loss percentage
15. Requisite resource requirements, staff capability, and feasibility of management control
16. Description of automated security features to be implemented
17. The impact of the proposed remediation on the assurance case
18. Description of the impact of a given remediation option on the business continuity strategy

Communicating with Authorization Decision-Makers

The body of evidence developed by the decision analysts is conveyed in an understandable fashion to the organizationally designated decision-maker who is responsible for authorization of the recommended responses. Therefore, the proper approach to ensuring the integrity and effectiveness of changes recommended for the ICT supply chain is to perform these activities as part of the set-up process:

1. Create an authorization control board composed of the appropriate decision-makers.
2. Alternatively, designate an appropriate management authority for approvals of change.
3. If the decision is below a threshold of significance, designate an authorizing agent.

The results of the analysis must be reported for whichever decision-making option is selected. The supporting documentation provides a full explanation of the change to be made and the implementation requirements for each remediation option and must clearly outline all relevant impacts for each option; it must be plainly understandable to lay decision-makers. The feasible remediation options are itemized in the report and must be expressed in a manner that is understandable to lay decision-makers. Each option recommended must be fully and clearly traceable to the assurance case.

Step 4: Manage and Oversee the Authorized Response

Practical management of the authorized remediation option calls for actions that are part of the larger context of the ICT SCRM process. Large-scale strategic risk management of any given supply chain requires coordination of the implementation and the eventual assurance of all authorized remediations. The policies, practices, tools, and regulations that guide and impact the specification and oversight of the authorized remediation should be well defined and understood across the organization. They should also be applied as a standard management approach within the everyday ICT supply chain product risk management operation.

Several generic types of responses fall under the heading of *remediations* in the conventional SCRM operation. The simplest and most utilitarian of those are the responses that have been designed and deployed to counter known threats and vulnerabilities. This activity is commonly termed 'patching.' The overall goal of ICT supply chain product risk management is to ensure the security and integrity of the developing product as it makes its way up the supply chain. The reality, though, is that a significant number of new vulnerabilities, which might threaten the security and integrity of the emerging product, will be discovered over the period of its development. The vulnerabilities might be found through investigations carried out by security professionals within the original equipment manufacturer, discovered by external organization product research, or reported by white-hat hackers. Even whistleblowers within an organization and/or any other interested party can point out vulnerabilities. On the darker side, they can also be revealed by the exploits of the black-hat community (Figure 6.6). At a minimum, the formal discovery process always entails the following stages:

1. *Detection*—A security vulnerability is identified and characterized.
2. *Warning*—The affected party(ies) are advised that a vulnerability or flaw has been found.
3. *Study*—The affected party(ies) verifies and validates the vulnerability's existence.
4. *Remediation*—The vendor develops a remediation option or patch.
5. *Release*—Authorized remediation is implemented, including requisite patches.

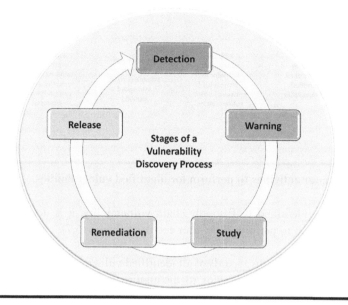

Figure 6.6   Stages of a vulnerability discovery process.

Responding to Known Vulnerabilities with Fixes

No matter the source, any vulnerability that has been discovered requires either patching or a more substantive form of risk mitigation. When a vulnerability is discovered, the discovering organization must report it to all parties that could be affected. The aim is to ensure that the specifics of the vulnerability are fully and completely known and explored by all the stakeholding organizations so that they can implement feasible risk mitigation options. The exploration process falls under the purview of step 2 "analyze reported vulnerability and understand operational impacts". However, the findings are ultimately coordinated and implemented in this step.

Responding to Known Vulnerabilities without Fixes

There might be operational justification for not fixing a vulnerability or for not fixing it as soon as possible. The justifications might include such obvious considerations as the lack of apparent harm, should the vulnerability be exploited. The most critical factor would simply be that the amount of time and resources required to implement the fix outweighed any potential cost should the threat occur. The supply chain oversight agent may lack the immediate resources to address the matter or the repair itself would be infeasible for business, technical, or operational reasons such as the unwillingness to take down a critical system. It is essential that every known vulnerability be formally monitored and managed. Therefore, in all

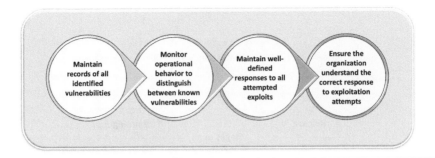

Figure 6.7   Four activities to perform for identified vulnerabilities.

instances of an identified ICT supply chain vulnerability, it is necessary for the management agent to perform these four activities (Figure 6.7):

1. Maintain continuous records of all identified vulnerabilities.
2. Monitor operational behavior to distinguish attempts to exploit a known vulnerability.
3. Maintain a well-defined response to any attempt to exploit a known vulnerability.
4. Ensure organization understands the correct response to an attempt at exploitation.

Fixing an Identified ICT Supply Chain Vulnerability

Once a decision has been made about whether to make a formal response, the agent performing the fix must understand all requirements and restrictions involved in making the change. Consequently, it is necessary for the overall response management function to establish a well-defined process to clearly convey all necessary technical and contextual specifications of the selected remediation to the change agent. In general, the agent making the change is embedded in the segment of the ICT supply chain that contains the vulnerability. This implies that the change agent is under control of the response management function but is not the same entity. Also, the change agent might be embedded in multiple organizations rather than a single setting. In both cases, individual or multiple, the response will be a substantive set of organizationally persistent controls, both electronic and behavioral.

Logically, to be effective, the controls must be implemented in a way that is uniformly standard across the ICT supply chain. The approach must guarantee consistent and disciplined execution of the process (Figure 6.8). Therefore, the recommended approach to implementing a specific fix is to

1. Identify the agent who will implement the fix—this may be supplier or integrator.

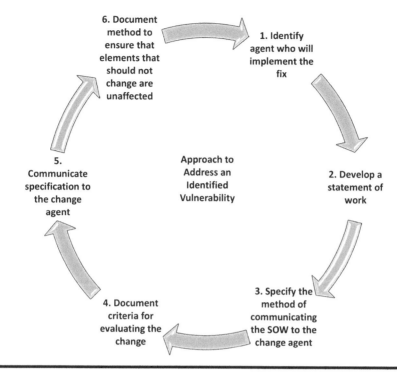

Figure 6.8 Approach to address an identified vulnerability.

2. Develop and document a Statement of Work (SOW) to achieve the intended outcome.
3. Specify the precise method for communicating the SOW to the change agent.
4. Document the criteria for evaluating the change to ensure successful remediation.
5. Communicate specifications to the change agent prior to instituting the change process.
6. Document method to ensure that elements that should not be changed are unaffected.

Step 5: Evaluate the Correctness and Effectiveness of the Implemented Response

Once a change to an ICT supply chain element has been made, it must be thoroughly validated to ensure that the target vulnerability has been addressed. This is particularly essential in ICT SCRM because it is an article of faith that in-stream changes to the product or process can create new problems. Accordingly, the hostile entities in the supply chain's ecosystem will typically examine any process or product remediations to identify new or different ways the changed item can be exploited. Consequently, the ICT supply chain response management function should continue to monitor and analyze the impact of a given change until its correctness is without question. This should be a reasonable period and is normally specified in the contract.

Some form of monitoring must occur after the change has been made; however, the overall aim is to confirm the long-term effectiveness of any alteration that has been made to the ICT supply chain process or product (Figure 6.9). Thus, it is good practice to do these six things:

1. Analyze the change in its everyday setting to identify any potential new points of attack.
2. Execute formal testing and reviews to identify any potential weaknesses or failures.
3. Ensure that all required integration and testing processes have been executed.
4. Ensure that all test results are reported to the appropriate authority.
5. Update the organization's understanding of overall supply chain architecture.
6. Modify the assurance case as appropriate.

Most remediations are implemented by agents that are internal to the selected element. The actual change takes place through the actions of another organizational process, such as the regular IT system management operation or the accounting

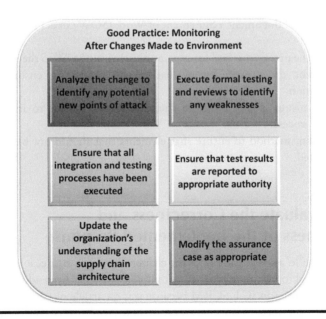

Figure 6.9   Good practice: Monitoring after changes made to environment.

function. The role of response management is to ensure that the change meets the criteria that have been stipulated in the statement of work or in the contract. This oversight function is a control process rather than a direct action and involves various review and testing methods (Figure 6.10). Therefore, it is considered good practice to perform these nine things:

1. Ensure that all assurance activities are conducted at their scheduled checkpoints.
2. Monitor the change action through joint reviews as specified in the SOW.
3. Ensure that any audit specified in the SOW is properly resourced and performed.
4. Ensure that any audit or review findings involving nonconcurrence are resolved.
5. Ensure that problems identified through joint reviews and audits are resolved.
6. Ensure that action items issuing out of each review are recorded for further action.
7. Ensure that closure criteria are specified and acted on.
8. Monitor changes to service levels as agreed to by contract.
9. Oversee the execution of change to ensure that required service levels are maintained.

The authority responsible for ensuring the accuracy of the eventual change should, in effect, close the account for that change. This is accomplished by ensuring that all the problem reports that are created during the correction process have been satisfactorily addressed and that the contractually specified closure criteria have been

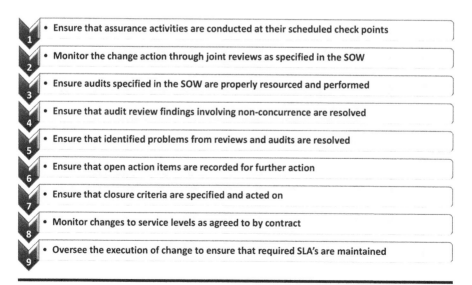

1. • Ensure that assurance activities are conducted at their scheduled check points
2. • Monitor the change action through joint reviews as specified in the SOW
3. • Ensure audits specified in the SOW are properly resourced and performed
4. • Ensure that audit review findings involving non-concurrence are resolved
5. • Ensure that identified problems from reviews and audits are resolved
6. • Ensure that open action items are recorded for further action
7. • Ensure that closure criteria are specified and acted on
8. • Monitor changes to service levels as agreed to by contract
9. • Oversee the execution of change to ensure that required SLA's are maintained

Figure 6.10   Oversight function good practice.

satisfied. This is typically ensured by the execution of a well-defined set of tests to ensure stipulated accuracy in accordance with good testing and evaluation practice.

The effective functioning of the change in the operational environment should be guaranteed by a formally defined validation and verification (V&V) program. This program must be able to ensure the correctness and integrity of the change as delivered. To accomplish this, the V&V function must be able to certify the functional completeness of the ICT supply chain operation against requirements. V&V must also be able to certify that the remediation has satisfied all function requirements stipulated in the formal statement of work. The V&V process itself compares any stipulated alterations against the technical description document and includes assurance of the physical completeness of the specified change to the supply chain process or system. This assurance must be documented using all appropriate technical descriptions.

Documenting the correctness and compliance of the remediation or change to the assurance case criteria requires an audit trail. The audit trail must provide objective, evidence-based assurance of the stipulated remediation option's correspondence with the change requirements that have been specified in the SOW and the contract. A variety of evidence might be relevant to assuring the correctness of the new change in light of assurance case criteria and includes evidence that these six things have been accomplished (Figure 6.11):

1. A formal SCRM remediation plan exists
2. Every remediation option has been reviewed and authorized
3. A schedule has been made for the work required to accomplish the change
4. Resource considerations are factored into the change authorizations

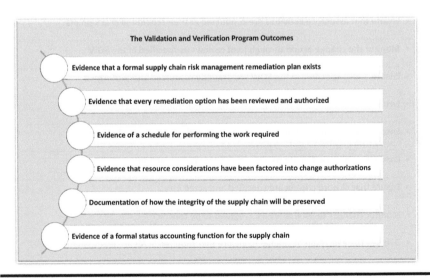

Figure 6.11   The validation and verification program outcomes.

5. Documentation has been made of how the integrity of the supply chain will be preserved
6. There is a formal status accounting function for the supply chain

Step 6: Assure the Integration of the Response into the Larger Supply Chain Process

All changes that are part of a remediation must be effectively reintegrated into the operational system. The decision-maker who authorized the remediation will eventually authorize the final reintegration stage. This approval is a formal organizational document and must be supported by the assurance evidence developed during the remediation change assurance outlined in step 5. Once the authorization is granted, the necessary physical alterations are made to reintegrate into the operational system of the elements of the supply chain that are involved. It is then necessary to conduct a technically rigorous process on all affected elements to ensure that the reintegration has been done correctly.

The reintegration process is also supported by a comprehensive V&V process. That process is specified and documented at the point where the designated remediation agent prepares the actual statement of work for making the necessary changes. The V&V program must certify that the reintegration of new components, patches, or process tasks is satisfactory and that all interfaces with unchanged areas of the supply chain are functioning properly (Figure 6.12). Given this requirement, it is considered good practice to

1. Confirm that the new supply chain configuration has been satisfactorily updated.
2. Certify reintegration correctness via an objective testing program.
3. Confirm reintegration maintains correct level of confidentiality, integrity, and availability.
4. Ensure documentation of the changed supply chain configuration is fully updated.
5. Ensure changed items maintain backward and forward traceability to prior states.
6. Ensure the configuration record is properly stored throughout the process.

In addition to certifying the accuracy of the altered supply chain entities, it is necessary to fully document the changed baseline of supply chain components. The current new operational configuration of the supply chain is maintained under strict management control. The documentation of that configuration is stored in an organizationally designated repository to ensure its integrity and confidentiality, if necessary.

In the case of large, global ICT product supply chains, it may be necessary to undertake a formal recertification or accreditation process to ensure trust among

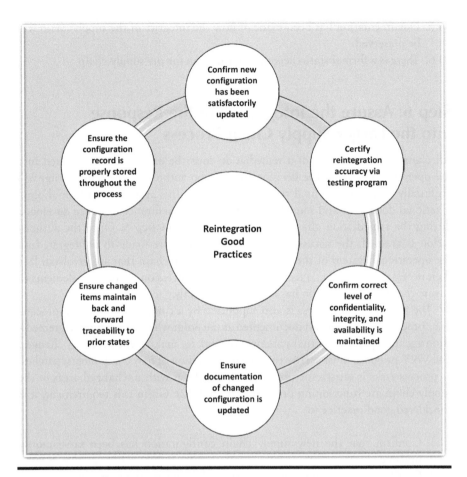

Figure 6.12   Reintegration good practices.

many suppliers. To ensure confidence in such a case, the supply chain must be audited by an appropriate third-party agent using a well-defined and commonly accepted auditing process. The findings of the audit process are typically accredited by formal certification, potentially as formal as the issuance of ISO 27000 accreditation or another third-party process mandated by the contract. In either case, formal assurance of the correct functioning of the supply chain should be done periodically to ensure continuing confidence of all participants and provided by a third-party agency that is considered trustworthy by all participants (ISO, 2013).

The audit requirements for any given engagement are normally established by contract or governmental regulation. Naturally, any assessments for formal certification of audited correctness must be performed by a properly certified lead auditor. This condition requires that the organization provide adequate resources and assurance to confirm the effective conduct of the audit process. Also, because trust is based on the objectivity of the audit, it is essential that the independence of the actual third party entity be guaranteed.

Confidence is enhanced by the adoption and use of a consistent, standard auditing methodology. Thus, the mutual acceptance of the auditing approach is an essential part of trust and must be assured. Where they are integral to the assurance of trust, audits should be performed on a basis timely enough to ensure continuing confidence. This ensures that the supply chain's security and integrity is controlled in a disciplined and secure fashion. It also ensures that the overall portfolio supply chain process will continue to meet its intended purpose.

Establishing a Supply Chain Assurance Infrastructure

The continuing assurance of trust up and down the supply chain involves formal organizational agreements to include risk analysis and response management that coordinate and sustain operational assurance. The agreements formalize the processes that the organization will employ to ensure that the assurance case and security architecture are reserved as intended and that the approved policies, processes, and methodologies for operational assurance, analysis, and response management are maintained correctly and securely.

The full supply chain architecture is the composite of all organizational and product components that are working together to produce a single integrated product entity. The purpose of a comprehensive architecture of components and processes is to help the organization provide rational identification, analysis, and authorized response to threats. Therefore, the security architecture must be holistic in focus and complete in its application. Establishing and maintaining a security architecture also ensures that the interrelationships among all elements of the supply chain are maintained correctly and effectively. Standard policies, processes, and methodologies are the meat and potatoes of the process. Standardization ensures that the organization's responses to threat are appropriate.

All responses must be flexible; it must be possible to alter them to ensure a continuously accurate response to alterations in the organization's threat environment. It is necessary to create a formal infrastructure to coordinate the policies, processes, and methodologies utilized to ensure ICT SCRM due to those processes and methodologies constituting the tangible elements of the SCRM process. By creating and documenting a tangible sustainment infrastructure, the policies, processes, and procedures also ensure continuous coordination between the overall supply chain management and the strategic risk management processes that ensure it.

A tangible security architecture involves the development, deployment, and continuous maintenance of the most appropriate and capable policies, procedures, security solutions, tools, and components. The aim of a security architecture is to maintain a dynamic security response to threats and changes as they occur in the

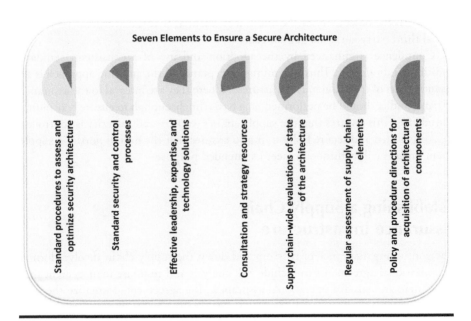

Figure 6.13   Seven elements for a secure architecture.

living supply chain (Figure 6.13). To accomplish this, the organization needs to plan, design, administer, and maintain these seven things:

1. Standard procedures to assess, integrate, and optimize security architecture
2. Standard security and control processes
3. Effective leadership, expertise, and technology solutions up and down the supply chain
4. Consultation and strategy resources to ensure effectiveness of the security architecture
5. Supply chain-wide evaluations of the continued effective state of the architecture
6. Regular assessment of supply chain elements to ensure security effectiveness
7. Policy and procedure directions for acquisition of architectural components

Policies for Operational Assurance: Method, Measurement, and Metrics

The success of an ICT SCRM operation requires the presence of appropriate policies, processes, and methodologies for operational assurance. The aim is to assure the security and integrity of supply chain elements up and down the framework that embodies them. Thus, an appropriate and capable set of policies, processes, and methodologies needs to be designed, documented, developed, deployed, and eventually sustained. The goal of operational assurance of the supply chain is to maintain an effective security and controls architecture that is both dynamic and resilient.

The most appropriate, tangible set of security methodologies, processes, and tasks should be developed and then deployed. To ensure this, the organization as a whole needs to embody and conduct a comprehensive and collaborative effort aimed at communicating the selected standard, commonly accepted security practices to all the participants in the supply chain process. This is usually achieved by the development of an organization-wide awareness and training program that effectively fosters the ideas and materials necessary to understand and follow good practice. Security and control awareness and knowledge of policies, procedures, tools, and standards must be championed and promoted. The process itself must be organizationally standard. Its coordination must be enforced by formal measures.

A supply chain cannot be controlled if the functioning of its processes is not sufficiently understood; to enable the necessary understanding, a security metrics program must be conducted and appropriate measures must be developed and collected. The metrics must be standard. They must allow the operational management function to optimize the ongoing security processes up and down the supply chain. Finally, these metrics must enable and be useful in the conduct of causal analysis.

Measures are applied through a measurement program that comprises formal teams whose purpose is to create and review objective data about the functioning of the supply chain. The teams must be established and coached in the best ways of applying the standard methodologies and processes for supply chain performance measurement. Where appropriate, the teams also assist with assessing compliance.

Metrics can be useful in improving the processes and practices of a given supply chain. The supply chain can be objectively evaluated using metrics and the efficiency, performance, and outcomes judged. Obviously, that requires a complete and unambiguous definition of the factors that will be used to undertake the analysis. Nevertheless, using metrics it is possible to model the individual elements and the infrastructure from a security perspective.

To minimize risks, the effective performance of the infrastructure needs to be fully understood in in terms of its overall operation. Metrics provide the evidence that informs and supports conclusions about the assurance case. A variety of evidence is relevant to creating and maintaining the assurance case in which the following are being accomplished:

1. Operational assurance supports achievement of strategic security requirements.
2. An adequately detailed operational assurance plan exists and is current.
3. Planning assumptions for current, known risks and threats have been documented.
4. Organization-wide standards and standard practices have been documented.

5. Technologies are documented along with the methodology for their secure operation.
6. All security updates have been verified, tested, and installed in a timely fashion.
7. A formal organization-wide security information sharing process exists.
8. The security information sharing process is updated as the security situation changes.

Building a Practical Supply Chain Sustainment Function

Sustainment defines and enforces integral management control over the operations of an ICT product supply chain. This type of management process is based on the selection and implementation of an explicit set of controls that are designed to achieve an intended outcome. At the heart of the sustainment process are the organizationally sanctioned actions that are designed to keep rational management control over the day-to-day activities of the supply chain. This is both a strategic and a local endeavor, and in the largest sense, operational control is a strategic activity designed to ensure the unqualified integrity of the supply chain. This level of control is a critical necessity because of the "weakest link" concern in the operation of any complex activity. However, the actual control can only be enforced at the local level due to each supply chain component's being an independent entity with its own practices and processes. Therefore, a single standard control process aimed at monitoring, error detection and correction, and cost reduction should be installed and operated at every level in the ICT product supply chain.

In concept, a single integrated control process applies to the organization and maintenance of responses to threat up and down the supply chain. The objective of this process is to ensure the integrity of local operations within the ICT product supply chain operation. Focusing management control of operations at the local level provides two primary advantages: it ensures the integrity of the component as built, based on the plans and intentions of the larger product development function, and it allows for the evaluation and performance of change to the local product, based on the integrity requirements of the larger ICT product supply chain operation.

If a single, formal standard process is imposed within every component, it also gives the ICT supply chain's top-level managers and policy makers direct input into the evolution of the product. It accomplishes this purpose by involving the appropriate set of top-level decision-makers in concrete decision-making about changes that are required for a lower level element of the product. Also, strategic imposition and enforcement of standard approaches to local operational changes provide a basis for objective measurement of security and quality of the supply chain as well as within the specific component. This makes testing and reviews easier, provides traceability of related components, and dramatically eases any change due to identified risks.

The actual day-to-day operational control process involves three major aspects of the routine control processes in an organization. Those three are: the risk identification process, the management control development process, and the organization-wide control performance measurement process, which is commonly called V&V. The latter two functions are the cornerstones of the overall operational control process because they shape the substantive response to identified risk. Therefore, they assure that the operational activities of a given component are maintained under acceptable management control.

From a practical strategy standpoint, three generic roles participate in the ICT product supply chain management control process: the *customer*; the *producer*, which can be either the supplier or the developer; and any associated *subcontractors*. The producer establishes a local operational management control plan in conjunction with the customer and ensures that it is understood and properly set up and maintained within the supplier organization. The producer appoints managers who have a defined responsibility for ensuring the control activities are properly executed, usually known as the company's supply chain risk managers. Besides enforcing local operational control over risks, the producer must be able to assure product security, ensured by audits that are carried out by designated third parties or local internal auditors.

The customer assigns a representative with authority commensurate to resolve all operational control issues pending between the producer and the customer. These might have to do with identified flaws in each ICT product supply chain component or involve planning of a response to new or identified risk. The customer representative has responsibility for approving proposals for additional cost or resource commitment. The role also ensures that the customer's organization observes all agreements between customer and supplier. The producer is generally responsible through the local control operation for insuring each subcontractor's uniform compliance, as specified for a given project. Aspects of the producer's agreement with the customer generally apply to the subcontractor; they are passed down to local control through the strategic management of the ICT product supply chain process at large.

Management control implies the existence of two standard and persistent processes. The first of these is the control over the operational activities of any given supply chain function. The second is the assurance of any operational activity authorized to address an identified or new risk. These are implemented through four generic, interdependent operational control processes aimed at ensuring an effective response to risk. In actual practice, these processes are tailored to the requirements of the local operation. These activities are as follows:

1. Risk analysis and response authorization
2. Verification and validation of the authorized response
3. Response integration and monitoring
4. Status accounting to verify compliance with specifications

Generic Management Roles

A set of generic management roles is responsible for executing the ICT product supply chain control process. The first of these is the local supply chain security operations control manager. This role ensures that the local requirements of the risk management process are properly carried out. The local supply chain security manager's overall goal is to manage the risk identification and analysis activity, manage all response authorizations, and verify that the selected response is both complete and properly executed.

The ICT product supply chain organization also defines and appoints a status accounting manager responsible for ensuring that all relevant elements of the supply chain are properly identified, accounted for, and maintained consistent with organizationally standard documentation. The primary responsibility of this individual is to establish and maintain a repository that baselines the authorized components of controlled items for each local entity in the ICT product supply chain. This role maintains detailed knowledge of the status of the product as built. To ensure the accuracy of this information the status accounting manager sets up and maintains an authorized product component repository. This repository constitutes the complete accounting of the product component for any element of the ICT product supply chain. Since items that are not in the ledger are, by definition, not part of the product, the status accounting manager is also responsible for keeping the repository up to date with any changes or new responses to risk. The status accounting manager maintains the components in the inventory for that supply chain element sufficient to reflect its current state. Therefore, only the status accounting manager is responsible for the authorization of entries into the product component repository.

The product assurance manager has the responsibility for ensuring that product integrity is maintained and that it is free of risk. The general role of the product assurance manager is to confirm that items in the organization's product component repository conform to the contractually authorized status of the part. This role also verifies the process and confirms the accuracy of changes that have been carried out to address risk. The product assurance manager also maintains documentation of all reviews conducted to support this process. In general, the product assurance manager must be able to guarantee that the components maintained in the repository reflect the intended state of the product at that point in time.

Conducting the Day-to-Day Operational Response Process

The goal of everyday management control over the ICT product supply chain process is the maintenance of product integrity. This probably goes without saying, but it is necessary to know exactly what constitutes the product to ensure that it is accurate. Therefore, all ICT product supply chain elements need to be identified, uniquely labeled, assessed, and documented in a repository of authorized components. These components are documented based on their interrelationships and dependencies within the ICT product supply chain. The documentation represents the basic architecture of the supply chain and its components. It must be noted here that the decisions that determine this architecture are made outside the operational management process, usually in some form of strategic development process. However, precise knowledge of the status of the elements of that architecture is an absolute prerequisite to managing it.

Once established, the repository of controlled components is maintained throughout the life cycle of the ICT product supply chain. Components defined at any level in this repository are maintained as such within the supply chain architecture. If a change is required for any element after it has entered the repository, then a formal process must be followed to move to a new representation of the product.

The management level authorized to approve a response must be explicitly defined. This is normally a strategic activity that takes place at the top of the ICT product supply chain. Authorization for approvals is always given at the highest practical level in the organization. This is done to ensure comprehensive acceptance of the changed architecture. Documentation of a change at any level in the basic architecture of the supply chain must be maintained at all levels.

The authorization process itself is hierarchical and composed of managers with sufficient authority to enforce the decision. At a minimum, there are three levels of control for an ICT product supply chain. The top level is composed of strategic policy makers for the entire supply chain.

There are generally two authorization entities at the local level, one for software risk management and the other for hardware risk management. The practical realization of the authorization process is managers who are the right level to understand all the ramifications of the risk and then oversee its remediation. The scope of who is authorized to approve what must be formally and explicitly defined, usually in the general strategic plan for the ICT product supply chain.

Response Management Process Planning

ICT product supply chain response management is defined and formally implemented through an operational management response plan. The entire supply chain's commitment to this plan must be rigorously enforced and maintained throughout the life cycle of a given ICT product supply chain. At the minimum, this plan should specify the three generic management roles: supply chain security manager, status accounting manager, and assurance manager, as well as the product component scheme that will be maintained in the repository and the composition of the various approval authorities. These managers also need to have their authority, scope, and responsibility for decision-making defined.

ICT product supply chain management control is a major organizational process; therefore, it requires a strategic plan. The plan is needed to assist management control of a substantive organizational activity. The plan describes the scope and application of the management control process, its activities, the procedures, and the schedule for performing all specified activities. Additionally, it states the roles and responsibilities of the individuals specifically responsible for performing each activity and the relationship of those roles to the rest of the organization.

The plan lays out the typical activities that constitute the three generic functions in the management control process, which must be stated in unambiguous terms. This includes the specification of procedures for performing all requisite control activities and can apply either to local elements of the ICT product supply chain or the supply chain as a monolithic entity. Because of its importance, the plan is generally a part of the overall supply chain contract. Its goal is to state the complete and correct mechanism that will be employed to perform the ICT product SCRM control process.

Deciding What to Secure

The essence of the process lies in establishing what to secure; therefore, a formal process for identifying and vetting the ICT supply chain components needs to be established to accurately characterize and prioritize the product entities that will be placed under management control for that project. Then the documentation that establishes the detailed status of each of these objects is identified for each of the components that are part of the entire ICT supply chain. The documentation must completely describe these control items. It must be kept in a repository sufficient to ensure that the product and its components are known and traceable at all points in the process. The actual elements that this repository documents are then kept under centralized management control for the duration of the product's life cycle.

Enforcing Management Control

The management control process is visible to the entire set of supply chain organizations, and this control should be centralized to enforce the requisite coordination. The harmonization of the ICT supply chain elements into a single managed entity has the greatest potential payoff for the organization. The actual analysis and authorization process lies at the heart of this coordination activity because analysis and approval ensure that the status of all elements included in the ICT supply chain is always known. As mentioned earlier, the control process responds to identified threat. It analyzes and evaluates the threat in terms of its potential impact, likelihood, and cost of remediation and passes that analysis along to the authorized decision-makers to either approve or disapprove the proposed response.

The decision is made by the authorized decision-maker in the management control process plan. It should be stressed that these decisions must be timely. Following receipt of authorization, the central coordinating entity that has been designated for the ICT supply chain provides a map of all affected elements along with a comprehensive resource analysis. This is given to the appropriate entity to develop or change the affected area of response. Normally, an entity other than the coordinating authority does the actual work to shape the response, which is typically described in a formal statement of work.

The work must produce an audit trail, which can be used to track the change to completion. The audit trail must be sufficient to ensure that accurate knowledge of the status of the organization's ICT supply chain is preserved at all times. The central management control process utilizes the audit function to record and track the evolution of the affected elements in the ICT supply chain. The central coordinating entity also provides a problem resolution service that allows any nonconcurrences between the ICT supply chain entities to be rationally resolved.

Finally, the centralized coordinating entity for the ICT supply chain is also responsible for ensuring that the status of all controlled elements in the supply chain is known and accounted for. In that respect, the coordinating entity maintains a comprehensive record of supply chain status and issues reports as scheduled, or required, to inform management of the status and history of any individual item in the supply chain.

Status Assessment

Status assessment is an important activity in the control of an ICT supply chain. Status assessment is normally conducted on a scheduled basis. However, it can occur at any point in the process if any type of management input is required. Status assessment is normally carried out by the product assurance management function. Thus, status reviews almost always take place at the end of any authorized response or change. The status review assures the functional completeness of the response or change against the statement of work as well as the initial analysis. Criteria for evaluation typically include whether (Figure 6.14)

1. The work has been properly authorized.
2. Stipulations of the statement of work have been satisfied.
3. The finished product addresses the identified threat.
4. The managerial and technical documentation have been properly updated.

Figure 6.14 Criteria for status evaluation.

Maintaining Documentation Integrity

In the end, the understanding of the practices embodied in the ICT supply chain is directly dependent on the integrity of the documentation that describes it. Therefore, master copies of the documentation that described the controlled elements of the supply chain should be stored in a secure place, with proper assurances of protection. This maintenance function is a lot more than simple filing because the amount of time and resources that went into creating a controlled supply chain environment will be lost if the integrity of the documentation is compromised. Therefore, the handling, storage, packaging, and delivery of status reports should be well defined and fully compliant with the strategic policies the organization has set for capable ICT supply chain management.

Chapter Summary

ICT products are acquired to facilitate organizational goals. In that respect, each product has a fundamental set of security requirements. The security environment is continuously changing. So, if some aspect of the evolving product does not meet those requirements, the organization needs to realign the product or its processes to bring them back into proper alignment. These factors and any others need to be accounted for in a real-world ICT product supply chain. The accounting should take place at both the overall supply chain system process level and within the processes of any given member organization within that sourcing chain. Consequently, a well-defined and properly stipulated process for managing change, both short and long terms, should be baked into any ICT product sourcing system.

In the largest sense, sustainment creates situational awareness for the supply chain. Sustainment monitors the hazard space and ensures that there is no present threat to the supply chain's ability to successfully deliver a contracted product or service. Sustainment identifies and records threats or problems, analyzes those problems, and takes the appropriate adaptive, perfective, or preventive corrective action to ensure that the assurance goals of the ICT product supply chain are being achieved. The goal of the sustainment process is to ensure unquestioned integrity and trust, up and down the ICT product supply chain. That assurance is underwritten by a disciplined, well-defined seven-step process.

Logically, the first step in that process requires stakeholders to maintain a continuous and accurate understanding of the threat space. Known threats can be countered. Threats that are unknown or unanticipated cause problems. Therefore, sustainment presupposes complete, accurate, and unambiguous knowledge of the current state of the ICT product supply chain at its most basic level of operation.

Then, the individual components, which are arrayed within that structure, are audited to characterize their state of compliance with contract requirements. Those individual compliance audits provide complete and fully documented assurance of the individual state of each component within the product supply chain tree. In that respect, the drilling down to each element of the tree structure provides the detailed status of the actual condition of the ICT product, at each relevant level in the supply chain. The process itself extends into the organization's operating environment as far as necessary to account for every potential attack vector.

The impacts of any unforeseen event or change to the ICT product supply chain must be fully understood to manage them intelligently. Threats can arise in the organizational ecosystem at any point in time and can represent a range of unanticipated outcomes. Therefore, the conjoined situational awareness and operational change analysis and management functions must be executed on a continuous, disciplined basis. Thus, it is a requisite of good practice that a well-defined operational analysis and response management function is created and continuously available to perform impact analyses and direct the requisite change management process to ensure the ongoing stability and security of the ICT supply chain.

In its most fundamental sense, supply chain sustainment is built around the process of change management. All changes that are made in response to the identification of new threats, or the need for different practice, must be authorized by the organizationally sanctioned decision authority. The approval for change should be based on an intelligent understanding of the costs and benefits of making it. Consequently, step 3 is fundamentally a policy support function in which capable people perform the necessary qualitative and quantitative analyses to support the decision.

So, once a report of a vulnerability or need for enhancement is identified, analysts evaluate the impact and likelihood of the potential harm along with the resource commitment required to make the change. Essentially, the projected resource commitment to effect a change is traded off against the cost of exploitation of the identified flaw. The trade-off analysis is necessary to ensure proper response.

Practical management of the authorized remediation option calls out actions that are part of the larger context of the ICT SCRM process. As we said earlier, large-scale strategic risk management of any given supply chain requires coordination of the implementation and the eventual assurance of all authorized remediations. The overall goal of ICT supply chain product risk management is to ensure the security and integrity of the developing product as it makes its way up the supply chain. The reality though, is that a significant number of new vulnerabilities, which might threaten the security and integrity of the emerging product, will be discovered over the period of its development.

Those vulnerabilities might be found through explicit investigations carried out by security professionals within the original equipment manufacturer. They might also be discovered by other organizations' product research or reported by white-hat hackers. Even internal whistleblowers or any other interested party can point out vulnerabilities. On the darker side, they can also be revealed by the exploits of the black-hat community.

Once a change to an ICT supply chain element has been made, it must be thoroughly validated to ensure that the target vulnerability has been addressed. This is particularly essential in ICT SCRM, because it is an article of faith that in-stream changes to the product or process can create new problems. Accordingly, hostile entities in the supply chain's ecosystem will typically examine any process or product remediations to identify new or different ways the changed item can be exploited.

All changes that are part of a remediation must be effectively reintegrated into the operational system. The decision-maker who authorized the remediation will eventually authorize the final reintegration stage. This approval is a formal organizational document and must be supported by the assurance evidence developed during step 5 "evaluate the correctness and effectiveness of the implemented response." Once the authorization is granted, the necessary physical alterations are made to reintegrate it into the operational system of the elements of the supply chain that are involved. It is then necessary to conduct a technically rigorous process on all affected elements to ensure that the reintegration has been done correctly.

The reintegration process is also supported by a comprehensive V&V process. That process is specified and documented at the point where the designated remediation agent prepares the actual statement of work for making the necessary changes. The V&V program must certify that the reintegration of new components, patches, or process tasks is satisfactory and that all interfaces with unchanged areas of the supply chain are functioning properly.

The continuing assurance of trust up and down the supply chain involves formal organizational agreements that coordinate and sustain operational assurance. That includes risk analysis and response management. These agreements formalize the processes that the organization will employ to ensure that the assurance case and security architecture are reserved as intended and that the approved policies, processes, and methodologies for operational assurance, analysis, and response management are maintained correctly and securely.

Key Concepts

- The SCRM sustainment process embodies three functions: response management, authorization, and process and product assurance.
- Sustainment is enabled by contract between the elements of the supply chain.
- The creation and documentation of changes and approvals is the basis for control.
- Supply chain sustainment management should be designed, coordinated, and enforced through a single controlling entity.
- Central coordination is ensured by standard, commonly accepted controls.
- Central coordination relies on complete knowledge of the status of all controlled elements in the supply chain.
- Testing and inspections are the means employed to ensure control status knowledge.

Key Terms

analysis: an explicit examination to determine the state of a given entity or requirement

assurance: the set of formal processes utilized to ensure confidence in a supply chain product

configuration management: rational control of change based on a formal process

control framework: a comprehensive set of standard behaviors intended to explicitly define all required processes, activities, and tasks for a given field or application

controls: activities built into a process designed to ensure a purpose

infrastructure: a planned entity with a consciously designated purpose

monitoring: specific oversight created by a planned collection and analysis of data

process: a collection of practices designed to achieve an explicit purpose

process design: the act of translating a given product development into specific steps along a timeline for an intended and verifiable outcome

repository: a collection of entities related by a similar purpose and timeframe

risk: a given threat with a known likelihood and impact

risk response: steps taken to reduce the impact of a given event

sourcing: the organizational process of product acquisition either through a development process or by the purchase of a commercial off-the-shelf solution

supply chain security: assurance that all elements in the supply chain are controlled and their present status is known

testing: validation of performance of a piece of software or system

transparency: full and complete understanding of the status of a product element, including the documentation set

verification and validation: the reviews and tests used to ensure a given set of criteria

References

International Standards Organization (ISO), ISO 27001, Information technology—Security techniques—Information security management systems—Requirements, ISO, 2013.

National Institute of Standards and Technology (NIST), NIST SP 800-53 Revision 4, Security and privacy controls for Federal Information Systems and Organizations, 2013.

Chapter 7

Building a Capable Supply Chain Operation

At the conclusion of this chapter, the reader will understand the following:

- The need for increasing supply chain risk management (SCRM) capability
- The stages and common features of capable SCRM
- The fundamental concepts of capability maturity models (CMMs)
- How the Software Acquisition (SA)-CMM is structured
- How to implement the SA-CMM

Introduction

The existence of a properly managed information and communication technology (ICT) supply chain is a critical requirement to leverage trust in an organization's sourced products. The formulation of a capable broad-scale management process relies on three commonsense principles:

1. Control the development and sustainment work using common best practice.
2. Adopt rigorous assurance practice at the component level.
3. Rationally plan for adverse contingencies.

A very large percentage of the counterfeiting, supply chain critical point-of-failure breakdowns, and capability concerns that were listed in the U.S. Government Accountability Office (GAO) report can be mitigated by simply ensuring that every entity up and down the supply chain adopts these simple management control practices (GAO, 2012). Strict control of the development and sustainment process for information and communication technology supply chain products ensures against unwanted functionality, while product defects are addressed by strict product assurance from the time of inception to the time of acceptance. When the inevitable supply chain failure does occur, there is a well-defined strategy in place to ensure that the problem is resolved in an intelligent fashion.

Control processes that satisfy these three principles are specified in a range of standard models of best practice. Because those standards are highly authoritative, their recommendations provide a coherent and logical high-level framework to ensure both the logic and relevance of all the elements of fundamental best practice they install. However, it is difficult to install a complicated set of process controls on a distributed supply chain in the real world without a well-defined and commonly accepted approach to accomplishing that task on a repeatable basis. This is the role of a *capability maturity model*. A standard CMM for information and communication technology supply chain management offers a systematic classification structure that will allow organizations to develop their capabilities for the management of existing and future supply chains.

The attractive feature of such a model is that the trust element is always enforced by a third-party audit. The standard and audited proof of capable, best practice underwrites two of the most important factors in global business: trust and competence. More specifically, a provably capable information and communication technology supply chain enhances the level of trust between supplier and customers and provides the customer with documented third-party assurance that all the suppliers and integrators in an information and communication technology supply chain are competent.

Obviously, a big part of ensuring trust relies on the ability of the supply chain's integrator and supplier organizations to guarantee that they can deliver a secure product that meets the contractual resource, timeline, and integrity criteria. The problem is that given the complexity of most information and communication technology supply chain products, it is difficult for any individual supplier to provide that sort of guarantee.

According to Watts Humphrey, three variables that provide the basis for trust in business are history, understanding, and awareness (1995). Information and communication technology supply chain organizations have a particularly difficult time assuring any of these three factors since the product could be produced in a distributed setting by any number of organizations that span the world. So, unless they have done business with each other before, companies often have little basis for gauging supplier performance. Additionally, even if two companies have experience with each other's work, there is no guarantee that a customer can rely on similar results given all the factors that impact supply chain operation: natural, manmade, or adversarial. A single, mutually agreed-upon model for ensuring uniformity in process must be available as a basis for independently assuring information and communication technology supply chain organizational competence. The model can then serve as the basis for building the commensurate trust.

The requirement invokes two practical conditions: First, the supplier must be demonstrably competent and second, the buyer must be able to reliably identify a competent supplier. If an acquirer has a long-standing history with a given supplier, that organization will know whether that supplier is competent. However, given distances and the global elements of business, this is not a common situation. As such, a defined process for independently proving and recognizing organizational capability is a necessity if information and communication technology supply chains are ever going to be established and managed in a rational fashion. This is the role of the formal capability maturity process discussed in this chapter.

Why a Capability Maturity Model?

Staged CMMs define ideal ways to enhance most types of human endeavor. They describe *what*, at a minimum, must be done to move from a state of incapability to one of optimized functioning as an organization. They do not specify *how* things must be done; they leave that definition to the individual organization. What they do specify are the processes that must be provably present in the operation and the required degree or level of capability of their execution.

The aim of the stages in a CMM is to establish order and a system for the way the organization goes about implementing optimum capability in the operation of a given information and communication technology product supply chain. The implementation sequence for a given set of management control practices is important because practices can be arranged to build on each other. Thus, the presence of basic assurance processes at one level can be leveraged into reporting and decision support activities at a higher level. But the foundational capability must come first.

A standard, staged, maturity model specifies what must be done to achieve a given level of capability. It also serves as a basis for obtaining audited assurance in which all participant organizations in an information and communication technology supply chain are at acceptably corresponding levels of capability. The process recommendations in a maturity model provide a template for setting up and running an effective capability maturity management process up and down the supply chain.

The experience of the Carnegie Mellon University Software Engineering Institute in developing the *Capability Maturity Model for Software* (SW-CMM) was directly applicable to developing the SA-CMM. The SW-CMM described the developer's (supplier's) role, while the SA-CMM described the customer's (acquirer's) role in the Acquisition process.

In the SA-CMM, an individual acquisition begins with the process of defining a system need. Some activities performed by the acquisition organization, such as planning, may predate the establishment of a project office. The SA-CMM includes certain precontract award activities, such as preparing the solicitation package, developing the initial set of requirements, and participating in source selection. In the SA-CMM, an individual acquisition ends when the contract for the products is concluded.

The SA-CMM identifies key process areas for four of its five levels of maturity. The key process areas state the goals that must be satisfied to achieve each level of maturity. In other words, progress is made in stages or steps. As shown in Figure 7.1, the five levels of maturity and their key process areas thus provide a road map for achieving higher levels of maturity.

A Staged Model for Increasing Capability in Supply Chain Management

The aim of all CMMs is to stipulate a well-defined and commonly accepted set of staged practices for the capable management of a desired set of processes. This is the aim of the Carnegie Mellon University Software Engineering Institute's, *SA-CMM* (Cooper, 2002). The SA-CMM specifies a standard set of organizational behaviors designed to ensure an increasingly capable acquisition operation. The model provides a staged maturity framework, along with specifying the actions that should be undertaken at each level to ensure a capable Acquisition process over time.

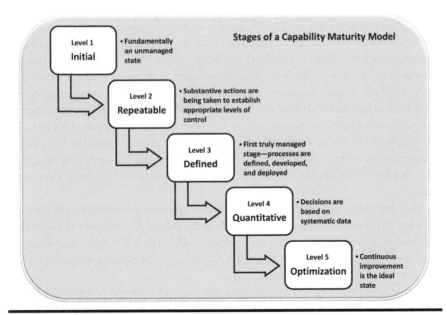

Figure 7.1   Stages of a capability maturity model.

The SA-CMM describes the activities that are essential to the effective management of an organization's information and communication technology product supply chain. As previously stated, earlier stages of the model are complementary, that is, they provide the foundation for later stages of the model and flow forward. Practices that are installed at a lower level do not need to be reinstalled at higher levels. The approach reduces unnecessary redundancy in subsequent key process areas and makes implementation more efficient (Cooper, 2002).

The SA-CMM is designed to be sufficiently generic that any organization, regardless of size, will be able to instantiate it. The model can be tailored to any context or security requirement of any customer organization in the supply chain. Also, the use of the SA-CMM is not limited to situations where the products are being acquired under formal contract (Cooper, 2002). The SA-CMM applies to the acquisition of all types of embedded and stand-alone applications, including commercial-off-the-shelf (COTS), either as a part of a system or separately (Cooper, 2002).

The SA-CMM is appropriate for use throughout all organizations in a supply chain. In the normal operation of an information and communication technology product supply chain, a wide range of products will be acquired from supplier to customer or integrator. The SA-CMM model is repeatedly applicable during the life cycle of a supply chain operation.

As shown in Figure 7.1, the SA-CMM defines five levels of maturity (Cooper, 2002). Each maturity level designates a level of process capability based on the presence of a set of well-defined key process areas. The six criteria that are used as the basis to characterize the achievement of each level of maturity are as follows (Cooper, 2002) (Figure 7.2):

1. *Goals*: Intended outcome of the implementation of a key process area
2. *Commitment to perform*: Organizationally standard policies to establish the process

Figure 7.2   Criteria to characterize achievement.

3. *Ability to perform*: Typically involves resources, organizational structures, and training
4. *Activities performed*: Planned procedures for performing and tracking the work
5. *Measurement:* Determine the status of the activities performed
6. *Verifying performance*: Ensure that the activities are performed as stipulated

Level One: The Initial Level

This is the fundamental unmanaged state. Thus, there are no key process areas at this level. This is the initial stage. It should probably be more appropriately labeled stage zero because all of the best practice activities of the Acquisition process are either undefined or applied ad hoc. For an organization to mature beyond the initial level, it must install basic management controls to instill self-discipline.

The overall Acquisition process is executed ad hoc, based on no specific plan. There is no commonly accepted or defined way of acquiring products. Staffing assignment is based strictly on availability. None of the staff are specifically trained in acquisition good practice, and there is no recognition that acquisition is a homogeneous, managed process. There are no top-level managers involved in or monitoring the actions of the Acquisition process.

Level Two: The Repeatable Level

Because the initial stage is essentially unmanaged, the Repeatable stage is the first point in the capability maturity process where substantive actions are being taken to establish an appropriate level of control over the information and communication technology Acquisition process. Figure 7.3 shows the focus areas for the Repeatable stage. The rudimentary control over the Acquisition process is enabled by a fundamental set of best-management practices. The goal of the practices at this level is to create, implement, and operate a formal set of strategic acquisition activities. These activities enable a process capable of monitoring the diverse customer, supplier, and integrator activities up and down the supply chain.

The primary benefit of the monitoring process is that it makes it possible to establish and track resources and coordinate schedules, as well as evaluate and accept the components of the information and communication technology product as they move up the supply chain ladder to the operational universe of the organization.

The key requirement at the Repeatable level is the installation of a foundational set of best practices. The practices allow the organization to leverage the fundamental processes needed for active management control, into a functional set of acquisition project plans. Consequently, every given acquisition project embodies a unified

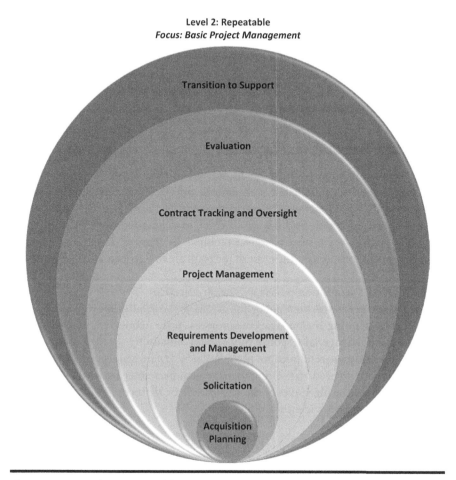

Figure 7.3   Level 2: Repeatable.

set of generic policies and standards that relate to its information and technology acquisitions into a standard operational program. The standardization allows for planning and for new projects to be based on lessons learned from similar projects. The aim of this set of unified policies is to make the contract management and the project management processes repeatable. That will ensure that best practices acquired from other similar projects are embodied in the planning for new ones.

At the Repeatable level, the organization has instituted all the necessary and fundamental acquisition management practices and controls. Acquisition managers work with suppliers and any subsupplier/integrators to establish direct and efficient lines of communication up and down the information and communication technology product supply chain. This allows the managers in the supply chain to specifically target and track the performance of the supplier organizations for their adherence to contractual requirements.

From a performance standpoint, organizations at level two embody the intentions stated in the label for that stage. They are "repeatable." This repeatability provides a stable basis for repeating earlier successes. For an organization to mature beyond this basic level of self-discipline, it must create and install a set of commonly accepted and well-defined standard management processes that can serve as a foundation for improvement.

Level Two: Acquisition Planning

Reasonable planning must first be conducted and all elements of the project considered, such as the budget, schedule, and strategic polity considerations, as well as the formulation of a set of functional requirements to define the shape of the actual acquisition. *Acquisition Planning* begins with a decision to acquire an information and communication technology system or service product. The process starts when a project team is formally constituted. The goals of the team are to (Cooper, 2002) (Figure 7.4)

■ Define the acquisition project's strategic objectives
■ Ensure there is a written policy for planning the acquisition
■ Ensure planning encompasses the project's entire Acquisition process
■ Establish managerial responsibility and accountability for the project
■ Ensure adequate resources are provided for software acquisition planning
■ Ensure experienced acquisition management personnel are available
■ Ensure that the objectives of the acquisition are defined
■ Enable formal risk identification

Figure 7.4   Goals of the acquisition planning team.

- Stipulate measurements to determine the progress of the acquisition
- Ensure acquisition activities are reviewed on a periodic basis

Level Two: Solicitation

The purpose of this standard common activity is to prepare a solicitation package that explicitly identifies the formal requirements of a given acquisition. *Solicitation* documentation allows the organization to select a supplier who is best capable of satisfying contract requirements. Solicitation undertakes all of the best practices crucial to the issuance of an effective solicitation package. It also prepares the process and criteria to evaluate responses and conduct supporting negotiations, leading to a recommendation for award of the contract. Solicitation ends with a contract award. The following outlines the solicitation process (Cooper, 2002) (Figure 7.5):

- Ensures that proposals are evaluated to ensure contract requirements
- Selects the supplier best qualified to satisfy the contract's requirements
- Ensures that a single decision authority is designated for each solicitation
- Ensures appropriate capability in the project team
- Ensures that groups supporting the solicitation understand solicitation objectives
- Ensures documentation includes specifications of requirements
- Ensures the statement of work includes detailed tasks
- Ensures contract acceptance procedures and criteria are known

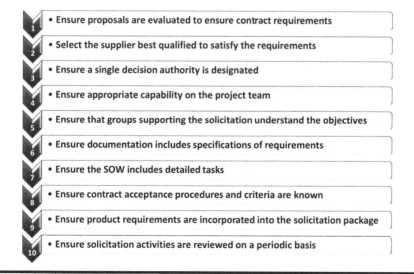

Solicitation Process for Acquisition

1. • Ensure proposals are evaluated to ensure contract requirements
2. • Select the supplier best qualified to satisfy the requirements
3. • Ensure a single decision authority is designated
4. • Ensure appropriate capability on the project team
5. • Ensure that groups supporting the solicitation understand the objectives
6. • Ensure documentation includes specifications of requirements
7. • Ensure the SOW includes detailed tasks
8. • Ensure contract acceptance procedures and criteria are known
9. • Ensure product requirements are incorporated into the solicitation package
10. • Ensure solicitation activities are reviewed on a periodic basis

Figure 7.5   Solicitation process for acquisition.

- Ensures product requirements are incorporated into the solicitation package
- Ensures solicitation activities are reviewed on a periodic basis

Level Two: Requirements Development and Management

This key process area addresses the development of contractual requirements and the subsequent management of these requirements for the duration of the acquisition. *Requirements Development and Management* begins with the translation of organizational requirements into a solicitation document containing specifications, and it ends with the transfer of responsibility for the support of the products.

Requirements Development and Management creates a common and unambiguous definition of the contractual requirements. These must be understood by the project team, end users, and the supplier team. Contractual requirements consist of both technical and nontechnical requirements and Requirements Management establishes and maintains agreement among the acquisition team regarding contractual requirements.

Requirements Management ensures that requirements are unambiguous, traceable, verifiable, documented, and controlled. The goal of Requirements Management is to ensure that (Cooper, 2002) (Figure 7.6)

- Contractual requirements are developed, managed, and maintained.
- Affected groups have input to the contractual requirements.
- Contractual requirements are traceable and verifiable.
- A requirements baseline is established prior to release of the solicitation package.

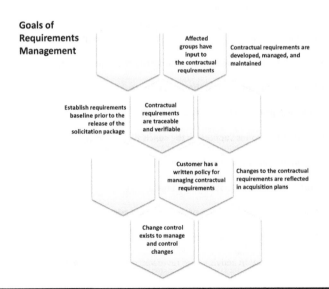

Figure 7.6   Goals of requirements management.

- The customer has a written policy for managing contractual requirements.
- Changes to the contractual requirements are reflected in acquisition plans.
- Change control exists to manage and control changes.

Level Two: Project Management

The purpose of *Project Management* is to manage the activities of the project management function to ensure a timely, efficient, and effective acquisition. Project Management activities involve planning, organizing, staffing, directing, and controlling project activities. Project Management begins when the project office is officially chartered and terminates when the acquisition is completed. The purpose of this key process area is to ensure the following (Cooper, 2002) (Figure 7.7):

- Project Management activities are planned, organized, controlled, and communicated.
- Problems discovered during the acquisition are managed and controlled.
- The customer reviews all project commitments.
- The project's acquisition plans are managed and controlled.
- Corrective action is undertaken when issues are discovered.
- Responsibility for Project Management is designated.
- A team that performs the project's acquisition management activities exists.
- Roles, responsibilities, and authority for the project functions are communicated.

Level Two: Project Management Activities

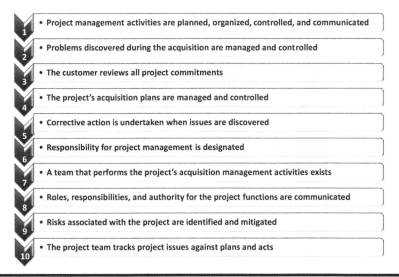

- 1 Project management activities are planned, organized, controlled, and communicated
- 2 Problems discovered during the acquisition are managed and controlled
- 3 The customer reviews all project commitments
- 4 The project's acquisition plans are managed and controlled
- 5 Corrective action is undertaken when issues are discovered
- 6 Responsibility for project management is designated
- 7 A team that performs the project's acquisition management activities exists
- 8 Roles, responsibilities, and authority for the project functions are communicated
- 9 Risks associated with the project are identified and mitigated
- 10 The project team tracks project issues against plans and acts

Figure 7.7   Level two: project management activities.

- Risks associated with the project are identified and mitigated.
- The project team tracks project issues against plans and acts.

Level Two: Contract Tracking and Oversight

The contract provides the binding agreement for establishing the requirements for the products to be acquired. It establishes the mechanism to allow the project team to oversee the supplier team's activities and evolving products and to evaluate any product being acquired. It also provides the vehicle for mutual understanding between the project team and the supplier of the requirements of the contract.

Contract Tracking and Oversight begins with the award and ends at the period of performance of the contract. The point of Contract Tracking and Oversight is to ensure that the complete set of contracted activities are being performed up and down the supply chain. Contract Tracking and Oversight also entails the identification of risks and potential vulnerabilities in the acquisition. Major functions in the Contract Tracking and Oversight process are as follows (Cooper, 2002) (Figure 7.8):

- Ensure that the effort is managed and controlled and complies with requirements.
- Ensure all contract commitments are agreed to and implemented by both parties.

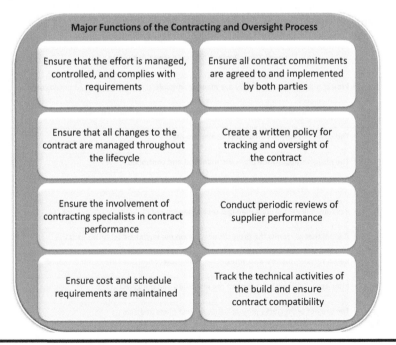

Major Functions of the Contracting and Oversight Process

Ensure that the effort is managed, controlled, and complies with requirements	Ensure all contract commitments are agreed to and implemented by both parties
Ensure that all changes to the contract are managed throughout the lifecycle	Create a written policy for tracking and oversight of the contract
Ensure the involvement of contracting specialists in contract performance	Conduct periodic reviews of supplier performance
Ensure cost and schedule requirements are maintained	Track the technical activities of the build and ensure contract compatibility

Figure 7.8 Major functions of the contracting and oversight process.

- Ensure that all changes to the contract are managed throughout the life cycle.
- Create a written policy for tracking and oversight of the contract.
- Ensure the involvement of contracting specialists in contract performance.
- Conduct periodic reviews of supplier performance.
- Ensure cost and schedule requirements are maintained.
- Track the technical activities of the build and ensure contract compatibility.

Level Two: Evaluation

The *Evaluation* key process area interacts with the *Solicitation*, *Requirements Development and Maintenance*, and *Contract Tracking and Oversight* key process areas. The Evaluation process starts with formal product requirements and ends when the acquisition is completed. The goal of the Evaluation key process area is to obtain sufficient objective evidence to prove that the activities and outcomes of the information and communication technology product supply chain fully fulfill all contractual requirements, prior to acceptance.

Evaluation involves the tailoring and documentation of the exact form of the information and communication technology product supply chain assessment approach. This includes the development of acceptance criteria, which are incorporated into both the solicitation package and the contract. Evaluations of the evolving products are conducted by both the supplier and project team. These are regularly carried out throughout the performance of the contract.

Outcomes of these evaluations are analyzed to determine product acceptability. Evaluation activities under this key process area are not the same as those performed by the supplier. The aim is to not duplicate the supplier's evaluation efforts while the product is under development. Evaluation supports the verification and establishment that the information and communication technology product meets contractual requirements. Major functions in the Evaluation process are as follows (Cooper, 2002):

- Develop evaluation requirements in accordance with contractual requirements.
- Plan and conduct evaluations throughout the total period of the acquisition life cycle.
- Create an empirical basis to support the product acceptance process.
- Ensure individuals performing evaluations have experience or receive training.

Level Two: Transition to Support

As the name implies, the *Transition to Support* key process area transitions the acquired information and communication technology product to operational use. Transition to Support begins with the earliest definition of requirements and ends when the product is operational in the customer organization. This process area ensures an effective and logical transition of the information and communication technology product from the supply chain to the customer organization. This process area also ensures the transparency of the software engineering and support environments.

The goal of Transition to Support is to facilitate the transition of the product of the information and communication technology supply chain, to the eventual customer organization. The support functions of the customer organization must be fully prepared to accept responsibility for the delivery of the finished products to ensure a smooth transition into the day-to-day operation. Thus, the customer function that will be responsible for transitioning the product is identified and embedded in the project's statement of requirements. Transition to Support also ensures the assurance that the information and communication technology product meets contractual requirements as delivered. Major functions in the Transition to Support process are as follows (Cooper, 2002) (Figure 7.9):

- Ensure the support organization has the capability to provide the required support.
- Ensure no loss in continuity of support to the transitioning products.
- Establish formal configuration management of the transitioning product.

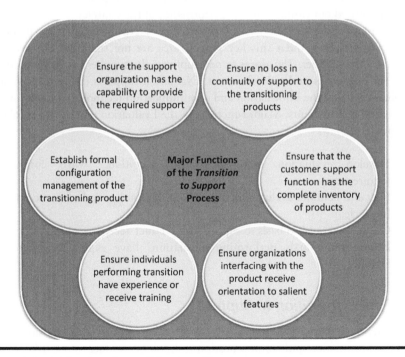

Figure 7.9 Major functions of the transition to support process.

■ Ensure that the customer support function has the complete inventory of products.
■ Ensure individuals performing transition have experience or receive training.
■ Ensure organizations interfacing with the product receive orientation to salient features.

Level Three: The Defined Level

The *Defined* level is the first truly managed stage in the five-stage maturity model. Figure 7.10 shows the focus areas for this level. The customer's formal, operational Acquisition process is fully established and includes processes for both contract

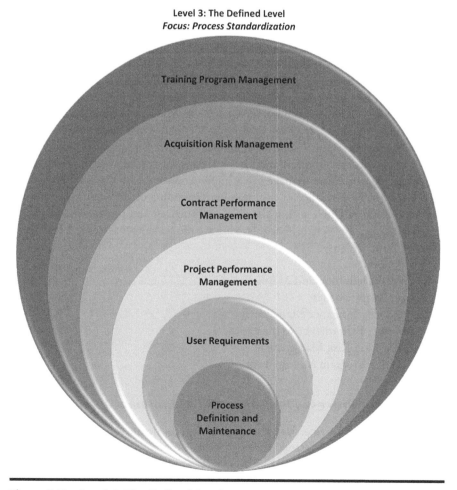

Figure 7.10   Level 3: The Defined Level.

management and project management. More importantly, the functions required to fulfill the intent of this process are integrated into the information and communication technology product supply chain for each product. This level could be more appropriately called the "well-defined and commonly accepted" level as the activities at Level Three comprise a standardized information and communication technology product supply chain assurance process. Level Three can be considered a "managed best practice" level in that there is an embedded process in place to facilitate and sustain the process definition and deployment process up and down the information and communication technology product supply chain.

The organizations in the supply chain utilize the customer organization's standard, defined, and established Acquisition process as a basis for defining the operational activities for a given information and communication technology product supply chain product development or COTS acquisition effort. Because the customer organization's standard Acquisition process is well defined and commonly understood, its managers have sufficient visibility into the work in progress within the project. Management and engineering activities are coherently integrated into the activities at each supplier level up and down the information and communication technology product supply chain. Level Three customer supply chain organizations can be said to be under management control in that all performance, cost, schedule, and requirement factors operate within the purview of a designated decision-maker and supply chain performance is monitored and tracked. Policies at the customer level ensure that the plans and contract requirements for any given supply chain product are monitored from a central point of control. That allows risks to be identified and managed in a highly coordinated fashion. This capability is leveraged by assurance of common understanding of well-defined processes, roles, and responsibilities within the given information and communication technology product supply chain. The following are the six key processes at Level Three (Cooper, 2002) (Figure 7.11):

1. Process Definition and Maintenance
2. User Requirements
3. Project Performance Management
4. Contract Performance Management
5. Acquisition Risk Management
6. Training Program Management

Level Three: Process Definition and Maintenance

The *Process Definition and Maintenance* key process defines and implements the customer organization's standard Acquisition process. Process Definition and Maintenance is focused on providing a correct and effective standard method for conducting the customer organization's information and communication technology product supply chain processes, as well as coordinating efforts to appraise and

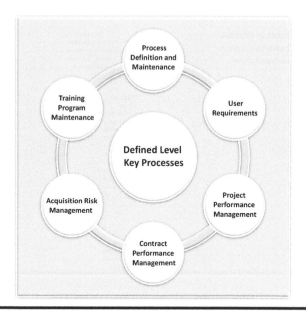

Figure 7.11 Defined level key processes.

improve the overall risk management of those processes. This key process area has operational responsibility for installing and sustaining that process in routine organizational functioning.

The key characteristic of this key process area is the establishment and maintenance of a process improvement group. Process improvement groups are standard implementation features in all SEI CMMs. These groups are formally established and have the mandated responsibility to define, sustain, and improve the organization's standard operational practices. They are also the basis for further leveraging the organization's capability maturity in their area of responsibility. That includes establishing guidelines for all information and communication technology product supply chain efforts as well as the actions to tailor the standard best practices from Levels Two and Three to each specific project situation. The goals of the Process Definition and Maintenance key area are to (Cooper, 2002) (Figure 7.12)

1. Define and control a standard Acquisition process for the customer organization
2. Coordinate process definition and maintenance activities across the organization
3. Collect, analyze, and make accessible, Acquisition process performance information
4. Create policy for Acquisition process definition and maintenance activities
5. Ensure the Acquisition process is defined and maintained in accordance with plans

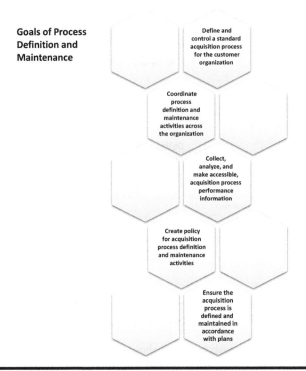

Figure 7.12   Goals of process definition and maintenance.

Level Three: User Requirements

Product requirements typically change over time, and if you are conducting a controlled information and communication technology product supply chain process, it is important to factor in the dynamic nature of the product environment into the operational control process to practically incorporate such changes into current and future versions of the products. The purpose of the *User Requirements* key process area is to elicit, analyze, translate, and communicate the evolution of the information and communication technology product requirements up and down the supply chain to ensure adequate understanding of up-to-date process and product needs.

User Requirements activities involve engaging customer organization representatives in ongoing dialogue with all elements of the information and communication technology product supply chain. This process elicits inputs into the way those requirements are developed and managed. The requirements then provide the basis for mutual understanding and agreement between the customer and all the elements of the supply chain. This key process area begins with the development of the initial set of requirements for the acquisition. It ends with the transfer of responsibility for the support of the acquired product from prime contractor to customer. The four goals of the User Requirements key process area are to (Cooper, 2002)

1. Obtain, document, and agree to customer requirements
2. Ensure customer requirements are accurately incorporated in the Acquisition process
3. Ensure that changes to customer requirements are managed as appropriate
4. Ensure that the customer has a written policy for performing requirement activities

Level Three: Project Performance Management

Project Performance Management ensures that the information and communication technology product supply chain is managed in accordance with the fundamental best practice definitions laid out at Level Two. The leveraging of capabilities established at one level by a coordinated management capability at a higher level is one of the fundamental characteristics of the Software Engineering Institute's CMM. The basic requirements for estimating, planning, and tracking an acquisition project are established in the Acquisition Planning, Project Management, and Contract Tracking and Oversight key process areas of Level Two.

The key process area anticipates problems and acts to mitigate risk. The emphasis in Project Performance Management at Level Three moves the customer organization from a reactive stance to a more proactive approach to problems and issues as they might be encountered in the information and communication technology product supply chain process. It accomplishes this by institutionalizing the customer organization's Acquisition process into a single unitary governance control system. Thus, the project management plans for any given acquisition are based on the organization's standard, well-defined Acquisition process. These plans describe how the project's defined Acquisition process will be implemented and managed. The four goals of the Project Performance Management key process area are to

1. Manage the project in accordance with a well-defined Acquisition process
2. Ensure the project achieves its acquisition objectives
3. Ensure that a written policy for performance management exists for each project
4. Ensure adequate training is provided to all acquisition project personnel

Level Three: Contract Performance Management

Contract Performance Management is the key process area that installs the requisite monitoring and control over the information and communication technology product supply chain. The activities in this process are oriented toward understanding and evaluating the specific contract performance of the suppliers and integrators in the supply chain process. This key process does systematic reviews of contract execution, and depending on the findings, adjustments can be made in supplier performance. Contract Performance Management begins when the contract is awarded and ends after the period of performance. The effect of instituting the Contract Performance Management key process is the implementation of a well-defined contract management function whose goal is to ensure that the evolving product and processes satisfy contract requirements.

This key process area is comprised of steps to evaluate products at contractually defined points, to judge whether the final product will satisfy contractual requirements. The focus of the Contract Performance Management process is to proactively manage suppliers and integrators to ensure that their performance complies with formal contract stipulations as well as resolve any issue that might arise from noncompliance. This key process is also an essential part of the project's risk management key process, in that it will identify potential weaknesses that might comprise areas of risk. The Contract Performance Management area builds on the Contract Tracking and Oversight and Evaluation key process areas at Level Two. The three goals of the Contract Performance Management key process area are to

1. Evaluate adequacy of supplier process, performance, and products
2. Foster a productive environment among stakeholders and suppliers
3. Ensure that there is a written policy for contract performance management

Level Three: Acquisition Risk Management

The aim of the *Acquisition Risk Management* key process area is to identify risks to the information and communication technology product supply chain as early as possible. It devises new and modified strategies to manage those risks. Acquisition Risk Management begins with the process of defining the acquisition needs and terminates when the acquisition is completed.

Acquisition Risk Management is planned through the Acquisition Risk Management Plan that details the processes to take place in acquisition planning and acquisition management. The Acquisition Risk Management Plan describes the management actions and procedures to identify, analyze, and rank risks and the risk handling methods to be applied. The goal is to embed risk management as an integral part of the acquisition organization's standard information and communication technology product supply chain process.

Acquisition Risk Management identifies the risks associated with the acquisition project with the optimum approach then decided based on the identified risks. Substantive well-defined and commonly accepted best-practice actions are undertaken to ensure against identified risks throughout the acquisition life cycle.

Risk identification includes the classification of risks based on determining the impact of each on quality, performance, schedule, and cost of a given information and communication technology product supply chain component. These risks are analyzed to determine whether adjustments are needed to acquisition strategy or engineering. The analysis includes determining root cause and impact of each risk area so that risk mitigations can be formulated and tested. Acquisition

Risk Management leverages the solicitation, project performance management, and contract performance management processes to do that. The five goals of the Acquisition Risk Management key process area are to (Cooper, 2002) (Figure 7.13)

1. Foster project-wide participation in identification and mitigation of risks
2. Ensure defined risk identification, analysis, and mitigation functions are performed
3. Conduct systematic reviews to judge the status of identified risks
4. Devise and implement risk mitigation plans for all supply chain elements
5. Conduct risk mitigation training to fit the specific acquisition

Level Three: Training Program Management

Training Program Management evaluates and devises training requirements that are tailored to the specific acquisition project in general and to each of the individual levels in the information and communication technology product supply chain. Members of the customer organization are oriented on the requirements of the Acquisition process, the specific project, and the skills and knowledge required. Training Program Management studies the proposed product and attempts to

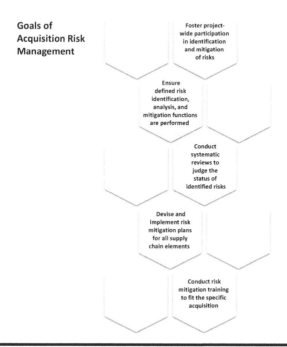

Figure 7.13 Goals of acquisition risk management.

define all necessary current and future skill set requirements. It also determines the mechanism for obtaining these skills.

Primarily, Training Program Management ensures the long-term sustainment of the skill sets and knowledge needed to ensure the proper execution of the information and communication technology product supply chain process. Some of these requirements can be satisfied by informal means. Other skill sets will require more formal training mechanisms. The appropriate vehicles are selected and used. This key process area is generally embodied in the common criteria *Ability to Perform*. The two goals of the Training Program Management key process area are as follows (Cooper, 2002):

1. Identify and provide the training required for capable project performance.
2. Ensure that the training program satisfies training needs.

Level Four: The Quantitative Level

This is the quantitative level of capability and because decisions are based on systematic data, this level can achieve a high degree of information and communication technology product supply chain security. The capability installed by quantitative management of the process fosters the customer organization's ability to operate its individual product supply chains within quantitatively measurable limits. This level of process capability allows an organization to empirically predict process and product assurance trends. When these limits are exceeded, action is taken to correct the situation.

This also serves to make the entire supply chain more effective and efficient since it will narrow variations in overall project performance to acceptable quantitative limits. Data on the defined Acquisition process and variations outside the acceptable quantitative boundaries are used to adjust the process to prevent recurrence of deficiencies. An acquisition organization-wide process repository provides for the collection and analysis of data from the projects' defined Acquisition processes. The customer organization defines quantitative policy objectives for management and assurance of its processes and products. There are two highly integrated process areas in this key process area:

1. Quantitative Process Management
2. Quantitative Acquisition Management

Level Four: Quantitative Process Management

Quantitative Process Management allows the organization to control its information and communication technology product supply chain activities using quantitative means that will ensure a more stable process because it is maintained under quantitative control. The customer organization collects performance data from its various information and communication technology product supply chain projects and uses that quantitative information to characterize the capability of the organizations and processes operating within the supply chain.

The capability data are, in turn, used by organizational management to set future performance objectives and to analyze the performance of their current processes. The capability data can also be used by the project teams to tailor the acquisition organization's standard Acquisition process to a given information and communication technology supply chain operation.

Quantitative Process Management requires the organization to set performance goals and measure performance against those goals as a means of making the necessary adjustments to keep supply chain performance within acceptable limits. Any deviation from performance goals is then measured and analyzed, and actions are devised to bring performance back within acceptable limits. The three general goals of the Quantitative Process Management key process area are to (Cooper, 2002)

1. Quantitatively measure and manage performance of each project
2. Characterize and quantitatively define the capability of the Acquisition process
3. Ensure that a policy for Quantitative Management is documented

Level Four: Quantitative Acquisition Management

The purpose of *Quantitative Acquisition Management* is to manage the contractual effort through the application of quantitative methods. As the customer organization progresses to higher levels of maturity, it will transition to the use of quantitative methods. This transition is a progression from Contract Performance Management, at Level Three, to an Acquisition Management approach that employs quantitative methods. This will leverage an increase in the security of the information and communication technology product supply chain.

Quantitative Acquisition Management uses the same process and product measurements as the Quantitative Process Management key process. It hinges on management review and empirical oversight. The actions of Quantitative Process Management at the project level and across the acquisition organization are integral to acquisition management. Thus, the Quantitative Acquisition Management process is integrated with the other key process, Quantitative Process Management. The goals of Quantitative Acquisition Management are to (Cooper, 2002)

1. Ensure the definition of quantitative measurement objectives for the acquired product
2. Quantitatively manage the activities of the contracted effort
3. Ensure that a Quantitative Acquisition Management plan is documented

Level Five: The Optimizing Level

Continuous improvement is the ideal state for an information and communication technology product supply chain operation. Level Five helps the customer organization achieve that state, and organizations are motivated to reduce the variations in performance that less capable supply chains will often experience. At the same time, they are constantly attempting to improve their overall level of performance.

The customer organization can detect processes that are likely candidates for optimization due to the empirical evidence that has been developed, which allows a Level Five customer organization to analyze each individual process for its effectiveness. The analysis can be used to refine policies. Technological innovations that exploit the best acquisition management and engineering practices can also be cataloged, assessed, and established. Improvements are leveraged from the advancements in performance that the supply chain process activities of Level Five provide, which facilitate effectiveness and efficiency up and down the supply chain.

Improvements are also fostered based on the use of any innovative new technology or techniques. The two Level Five key process areas are as follows (Cooper, 2002):

1. Continuous Process Improvement
2. Acquisition Innovation Management

Level Five: Continuous Process Improvement

The *Continuous Process Improvement process* is goal based and centers on the definition of a set of realistic quantitatively assessable process improvement objectives for the information and communication technology product supply chain. This is an ongoing effort with the aim to proactively and systematically identify, appraise, and implement substantive improvements within the information and communication technology supply chain operation, as well as integral processes up and down the supply chain.

The overall goal of Continuous Process Improvement is to leverage the effectiveness of the Acquisition processes that have been established within the organization by means of the first four levels of capability maturity. The Process Definition and Maintenance key process area defines the actions for implementing improvements to the customer organization's standard information and communication technology product supply chain operation. The Project Performance Management key process area establishes the actions for making the necessary changes to the targeted areas for improvement. Then, the Level Three Acquisition Innovation Management key process area defines the actions required to adapt and transform the proposed new techniques and technologies into the revised operation.

The targets of the improvement function are defined by quantitatively stated goals and objectives as well as the tailored Acquisition processes for each project. Level Five ensures that those processes are optimized by means of a formally managed Continuous Process Improvement process. The commitment to Continuous Process Improvement must be maintained organization-wide. Improvement opportunities are best identified by the people doing the work. Therefore, a lot of the implementation of this key process is based on training and incentive programs, which are established to ensure that all personnel are involved in the Continuous Process Improvement activities. The two goals of Continuous Process Improvement are to (Cooper, 2002)

1. Ensure that involvement in Continuous Process Improvement is organization-wide
2. Continuously improve the customer organization's Acquisition process

Level Five: Acquisition Innovation Management

Acquisition Innovation Management is the practical element of continuous improvement. It appraises, implements, adopts, and transfers new techniques and technologies in support of the overall information and communication technology product supply chain operation and its constituent processes. The focus of this key process is on improving the customer organization's practical Acquisition process through the adoption and implementation of new techniques and technologies. Adoption of these new techniques and technologies is carried out in conjunction with the mission and objectives of the Continuous Process Improvement key process area that was just discussed.

Acquisition Innovation Management's only aim is to underwrite the continuous improvement of the customer organization's product supply chain operation. It works to continually improve the customer organization's Acquisition process through the adoption and transfer of new techniques and technologies. Acquisition Innovation Management is a universal goal of the Level Five acquisition organization. Nevertheless, the practical innovation adoption activity will inevitably be different in every individual acquisition project. Therefore, each supply chain project can be considered a novel application of this management process. Yet, the environment for innovation and optimization should be fostered and led by the acquisition organization.

The actual identification and management of new or emerging techniques and technologies comprises a product space that is as wide as the field itself. These techniques and technologies might be novel or even over the horizon, and they can range from introduction of a new technology to carry out mundane project purposes, such as automated configuration management support tools. Or it can involve improvements to management strategies and across-the-board policies throughout the information and communication technology product supply chain operation. The goals of the Acquisition Innovation Management key process area are to (Cooper, 2002)

1. Proactively improve the Acquisition Process and Acquisition Management
2. Ensure organization-wide involvement in Acquisition Innovation Management

Practical Evaluation of Supply Chain Process Maturity

The capability maturity process is established through assessments. These assessments provide the basis for deciding the maturity level of each information and communication technology product supply chain operation. The capability assessments are extremely useful in the selection of the organizations that will comprise the components of any given product supply chain or evaluating the ongoing effectiveness of a supply chain that underlies a COTS product due to capability assessments focusing on pinning down any, and all, risks associated with a given supplier. A top-to-bottom capability evaluation of every component organization in an information and communication technology product supply chain operation might be too costly to perform. Not every situation warrants such an expense; however, capability assessments are necessary at the top levels in the supply chain hierarchy, especially where important contracts are being bid.

Since information and communication technology product supply chain integrity is at stake, a capability assessment tends to look like an audit. These types of audits are powerful tools since they are based on documentation and evidence, not judgment. Therefore, SA-CMM capability evaluations ought to be dropped on bidders in the supply chain hierarchy as part of the normal supply chain formulation process. These types of evaluations may also involve the potential subcontractors to monitor and assure the ongoing performance of lower-tier organizations in the supply chain.

At whatever tier, the supplier and integrator attributes considered in a capability assessment can be factored into two assessment targets: (1) the key process areas and (2) additional situational or contractor attributes, which are typically considered given any special requirements for the product and are situation-specific. In practice, several potential contractors should be evaluated for a specific supply chain spot and the findings related to requisite capabilities or anticipated risks factored into the decision to let the bid.

What follows is a short summary of the typical steps involved in these two types of assessments. The first step in both cases is to put together a formal assessment team. The members of the team should be professionals knowledgeable in information and communication technology acquisitions and supply chain product risk management. Obviously, this team should be intimately familiar with the fundamental concepts of the assessment model as well as be up to speed on the particulars of the SA-CMM auditing process. Generally, teams are formulated to investigate a specific key process area. The actual assessment is dictated by a set of checklist cover sheets. These checklists guide the assessment team during the appraisal and should be used in a standard fashion throughout the process. The checklist states the goals for the key process area, lists the activities that must be observed and rated, and describes each of the common features related to organizational commitment such as staffing, resourcing, and formal management commitments.

Where a large, complex, typically global, information and communication technology product supply chain is concerned, there is an intermediate step, which is simply to agree on what encompasses the entities in the supply chain. The organizational context must be unambiguously clear and understood for an evaluation across a range of cultures to be properly targeted and in that respect considered valid. Depending on the mission, goals, and context, the assessment target could be the entire organization, a division within that organization, a product line within that division, or even a logical collection of projects. Even if the organization is housed in one location and not diversified in terms of products, there is a need to determine what is being assessed.

The appraisal begins before the actual on-site assessment takes place. The appraisal team and the team members from the organization being appraised seek to acquire as much information as they can about the organization that will be assessed. This must be driven by a judiciously designed and standard questionnaire. The questionnaire asks the organization about their specific approach to satisfying the operational requirements of the appraisal. It records the number of employees, the key process areas, the target maturity levels for those processes, and the business domains in which they apply. Examples of information requested are as follows: the product, or project name, how many people are currently working on it, when it began, when it is scheduled to end, and the current stage of the project. The areas surveyed are the appropriate SA-CMM key process areas for the desired level of capability maturity. Once this questionnaire has been answered, the assessment team does a simple gap analysis to determine which areas are being adequately executed and which are not being satisfactorily performed. There may be instances where there is insufficient information. If that is the case, then additional targets for investigation, methods, tools, technologies, or criteria may be developed to support the assurance of conformity to the stipulations of the selected key process area.

The next step in the process is the on-site visit to the site being assessed. Generally, the team reviews the target documentation identified in the first step. The assessment team conducts a series of interviews to get a feel for the status of their assigned key process area at that site. After meeting with all the necessary, pertinent individuals and reviewing the targeted documentation, the team decides whether their target of evaluation satisfies the goals and requirements of the key process area they are investigating. Generally, the team is required to quantitatively document its rationale for whatever judgments it makes.

Following the site visit, the assessment team produces a list of findings. This list identifies the areas of compliance and the strengths and weaknesses of that specific agreement area. The findings themselves have different purposes depending on the type of evaluation being done. If it is a prime contractor, e.g., the supplier organization, the assessment conclusions serve as a basis for the recommendations with respect to capability of the organization to supply a trustworthy information, or communication technology product component. If this is an evaluation of the entire supply chain capability, then the results of this assessment are amalgamated into a judgment about the anticipated risk of dealing with the supply chain as a whole. This process usually terminates in a set of formal recommendations to contract with the supplier organization for products or services.

Maturity Rating Schemes

The primary goal of a CMM-based assessment is to decide the maturity level of an individual supply chain organization or the supply chain. In general, this requirement is satisfied by consolidating all relevant assessment information into a single basis for judgment and then applying the rating criteria suitable to the appropriate level of the SA-CMM. As we have seen, CMM is made up of a hierarchy of key process areas and component practices that are leveraged by areas at a lower level of maturity in the model. Based on the assessment results, each of these can be rated as follows:

- *Satisfied*: The component or method that satisfies the goals of the process is in place.
- *Unsatisfied*: Valid weaknesses are identified that significantly impact goals.
- *Not applicable:* The component does not apply in an organization's environment.
- *Not rated:* The component is outside the scope of the assessment.

For the activities in a key process area to be correct, all of the best practices that it comprises must be satisfied or not applicable. Each maturity level contains several key processes that must be satisfied. For example, the Level Four key processes are quantitatively manage the project and quantitatively manage the acquisition. To be considered a Level Four organization, the two key process areas must be fully satisfied. The maturity level rating is presented in the final report to the requestor. Also as a part of the final report, a detailed outline of where the organization stands for each key process area is presented. Finally, the organizational strengths and weaknesses are summarized.

Objective evidence must be used to decide whether an organization complies with a certain Key Process. Documents and interviews are used to decide this. Documents could include copies of policies and procedures, code libraries, electronic records, and visual media. Two levels of documents are reviewed. The first of these are organization-level documents. These express the practices that every member of the organization should know, understand, and use. Organizational documents might include the following (Cooper, 2002):

1. Organizational size and costing procedures
2. Standard reporting practices required across the organization
3. Standard metrics required for projects
4. Tailoring guidelines and waiver procedures
5. Training plans for the organization
6. Policies, procedures, and standards for engineering
7. Standard life cycle activities such as design, programming, and testing
8. Policies, procedures, and standards for support activities

Since the information and communication technology SCRM process is performed at the individual organizational level, as well as the supply chain as a whole, there are relevant individual organization-focused documents that must be included in the general assessment. These organization-specific documents are necessary to better understand and define the activities needed to coordinate and integrate the engineering activities for a specific component in the supply chain, at a given level in the overall process. These documents specify the day-to-day activities that are undertaken for a target for assessment within the supply chain. Among other things, individual organizational level documents can include the following (Cooper, 2002):

- Project status reports and schedules
- Configuration management change requests
- Test records
- Training records
- Historical data derived by comparing planned versus actual trends

At the end of the assessment, the findings, nonconformities, and other observations are compiled into a report. The elements of this report include the following:

- The scope and objectives of the assessment
- Details of the assessment program including team members and assessment dates
- Copies of nonconformity reports
- The team's recommendations for each target for evaluation

Chapter Summary

A properly managed information and communication technology supply chain is a critical requirement to leverage trust in an organization's sourced products.

However, it is difficult to install a complicated set of process controls on a distributed supply chain in the real world without a well-defined and commonly accepted approach to accomplishing that task on a repeatable basis. This is the role of a CMM. A standard CMM for information and communication technology supply chain management offers a systematic classification structure that will allow organizations to develop their capabilities for the management of existing and future supply chains.

Obviously, a big part of ensuring trust relies on the ability of the supply chain's integrator and supplier organizations to guarantee that they can deliver a secure product that meets the contractual resource, timeline, and integrity criteria. The problem is that, given the complexity of most information and communication technology supply chain products, it is difficult for any individual supplier to provide that sort of guarantee.

Staged CMMs define ideal ways to enhance most types of human endeavor. They describe what, at a minimum, must be done to move from a state of incapability to one of optimized functioning as an organization. They do not specify how things must be done. They leave the definition of that to the individual organization. What they do specify are the processes that must be provably present in the operation and the required degree or level of capability of their execution.

The aim of the stages in a CMM is to establish order, or a system, for the way the organization goes about implementing optimum capability in the operation of a given information and communication technology product supply chain. The implementation sequence for a given set of management control practices is important because practices can be arranged to build on each other. Thus, the presence of basic assurance processes at one level can be leveraged into reporting and decision support activities at a higher level. But the foundational capability must come first.

A standard, staged, maturity model specifies what must be done to achieve a given level of capability. It also serves as a basis for obtaining audited assurance that all the participant organizations in an information and communication technology supply chain are at acceptably corresponding levels of capability. The process recommendations in a maturity model just provide a template for setting up and running an effective capability maturity management process up and down the supply chain.

The Software Acquisition Capability Maturity Model (SA-CMM) identifies key process areas for four of its five levels of maturity. The key process areas state the goals that must be satisfied to achieve each level of maturity. In other words, progress is made in stages or steps. The levels of maturity and their key process areas thus provide a road map for achieving higher levels of maturity.

The first of these is the Initial Level. This is really the fundamental unmanaged state. Thus, there are no key process areas at this level. This is the initial stage. It would be more appropriately labeled stage zero because all the best practice activities of the Acquisition process are either undefined or applied ad hoc. For an organization to mature beyond the initial level, it must install basic management controls to instill self-discipline.

Level Two is the Repeatable Level. Because the initial stage is essentially unmanaged, the Repeatable stage is the first point in the capability-maturity process where substantive actions are being taken to establish an appropriate level of control over the information and communication technology Acquisition process. The rudimentary control over the Acquisition process is enabled by a fundamental set of best-management practices.

The goal of the practices at this level is to create, implement, and operate a formal set of strategic acquisition activities. These activities enable a process that is capable of monitoring the diverse customer, supplier, and integrator activities up and down the supply chain. The primary benefit of the monitoring process is that it makes it possible to establish and track resources, coordinate schedules, and evaluate and accept the components of the information and communication technology product as they move up the supply chain ladder.

From a performance standpoint, organizations at Level Two embody the intentions stated in the label for that stage. They are "repeatable." That repeatability provides a stable basis for repeating earlier successes. Nevertheless, for an organization to mature beyond this basic level of self-discipline, it must create and install a set of commonly accepted and well-defined standard management processes that can serve as a foundation for improvement. These management practices are as follows (Cooper, 2002):

1. Acquisition Planning
2. Solicitation
3. Requirements Development and Management
4. Project Management
5. Contract Tracking and Oversight
6. Evaluation
7. Transition to Support

Level Three: The Defined Level is the first truly managed stage. The customer's formal, operational Acquisition process is fully established here. The Defined Level includes processes for both contract management and project management. More important, the functions required to fulfill the intent of this process are integrated into the information and communication technology product supply chain for each product.

This level would be more appropriately called the "well-defined and commonly accepted" level, as the activities at Level Three comprise a standardized information and communication technology product supply chain assurance process. Level Three can be considered a "managed best practice" level in that there is an embedded process in place to facilitate and sustain the process definition and deployment process up and down the information and communication technology product supply chain. The following are the six key processes at Level Three (Cooper, 2002):

1. Process Definition and Maintenance
2. User Requirements
3. Project Performance Management
4. Contract Performance Management
5. Acquisition Risk Management
6. Training Program Management

Level Four: The Quantitative Level is the next level of capability. Because decisions are based on systematic data, this level can achieve a high degree of information and communication technology product supply chain security. The capability installed by quantitative management of the process fosters the customer organization's ability to operate its individual product supply chains within quantitatively measurable limits. This level of process capability allows an organization to empirically predict process and product assurance trends. When these limits are exceeded, action is taken to correct the situation.

This also serves to make the entire supply chain more effective and efficient since it will narrow variations in overall project performance to acceptable quantitative limits. Data on the defined Acquisition process and variations outside the acceptable quantitative boundaries are used to adjust the process to prevent recurrence of deficiencies. An acquisition organization-wide process repository provides for the collection and analysis of data from the projects' defined Acquisition processes. The customer organization defines quantitative policy objectives for management and assurance of its processes and products. There are two highly integrated process areas in this key process area:

1. Quantitative Process Management
2. Quantitative Acquisition Management

Level Five: The Optimizing Level is the ideal state for an information and communication technology product supply chain operation. Level Five helps the customer organization achieve that state. Level Five organizations are motivated to reduce the variations in performance that less-capable supply chains will often experience. At the same time, they are constantly attempting to improve their overall level of performance.

The customer organization can detect processes that are likely candidates for optimization. That is because the empirical evidence has been developed that allows a Level Five customer organization to analyze each individual process for its effectiveness. That analysis can be used to refine policies. Technological innovations that exploit the best acquisition management and engineering practices can also be cataloged, assessed, and established. Improvements are leveraged from the advancements in performance that the supply chain process activities of Level Five provide, which facilitate effectiveness and efficiency up and down the supply chain.

Improvements are also fostered based on the use of innovative technology or techniques. The two Level Five key process areas are (Cooper, 2002):

1. Continuous Process Improvement
2. Acquisition Innovation Management

Assessment: The capability maturity process is established and validated through *assessments*, which provide the basis for deciding the maturity level of each information and communication technology product supply chain operation. These capability assessments are extremely useful in the selection of the organizations that will comprise the components of any given product supply chain or the evaluation of the ongoing effectiveness of a supply chain that underlies a COTS product. That is because capability assessments focus on pinning down any, and all, risks associated with a given supplier. A top-to-bottom capability evaluation of every component organization in an information and communication technology product supply chain operation might not be warranted in every situation. But capability assessments are necessary at the top levels in the supply chain hierarchy, especially where important contracts are being bid.

The first step in both cases is to put together a formal assessment team made up of professionals knowledgeable in information and communication technology acquisitions and supply chain product risk management. Obviously, this team should be intimately familiar with the fundamental concepts of the assessment model as well as be up to speed on the particulars of the SA-CMM auditing process. Generally, teams are formulated to investigate a specific key process area. The actual assessment is dictated by a set of checklist cover sheets. These checklists guide the assessment team during the appraisal and should be used in a standard fashion throughout the process. The checklist states the goals for the key process area, lists the activities that must be observed and rated, and describes each of the common features related to organizational commitment such as staffing, resourcing, and formal management commitments.

Where a large, complex, typically global, information and communication technology product supply chain is concerned, there is an intermediate step, which is simply to agree on what encompasses the entities in the supply chain. Because the organizational context must be unambiguously clear and understood for an evaluation across a range of cultures to be properly targeted and in that respect valid. Depending on mission, goals, and context, the assessment target could be the whole organization, a division within that organization, a product line within that division, or even a logical collection of projects. Even if the organization is housed in one spot and not diversified in terms of products, there is a need to determine what is being assessed.

The primary goal of a CMM-based assessment is to decide the maturity level of an individual supply chain organization, or the supply chain. In general, this requirement is satisfied by consolidating all relevant assessment information into a single basis for judgment and then applying the rating criteria suitable to the level of the SA-CMM. As we have seen, CMM is made up of a hierarchy of key process areas and component practices that are leveraged by areas at a lower level of maturity in the model. Based on the assessment results, each of these can be rated as follows:

- *Satisfied*: The component or method that satisfies the goals of the process is in place.
- *Unsatisfied*: Valid weaknesses are identified that significantly impact goals.
- *Not applicable:* The component does not apply in an organization's environment.
- *Not rated:* The component is outside the scope of the assessment.

Key Terms

assurance: the set of formal processes utilized to ensure confidence in a supply chain

baseline security: a minimum level of acceptable assurance of proper performance

common features: process characteristics designed to measure correct execution

control frameworks: large strategic collection of controls array to achieve a purpose

controls: behaviors built into a product supply chain to ensure a secure state

key practices: formal practices to ensure that the work is correctly executed

key processes: operations that are needed for and indicative of good practice

process architecture: the method of organization of the overall supply chain or SDLC

process maturity: the level of capability of a process based on routine key practices

process specifications: the explicit work rules and requirements of a given operation

risk assessment: the evaluation of the likelihood and impact of a given threat

risk mitigation: steps taken to reduce the impact of a given event

security controls: explicitly designated behaviors to ensure proper performance

system development life cycle (SDLC): a formal series of steps in a process

testing and evaluation: validation of the performance of the assessment target

vulnerabilities: explicit known weakness that can be exploited by a given threat

References

Cooper, J. and Fisher, M., Software acquisition capability maturity model (SA-CMM) Version 1.03, Software Engineering Institute, Carnegie Mellon University, Pittsburgh, PA, 2002.

Humphrey, W., *A Discipline for Software Engineering*, Addison-Wesley, Reading, MA, 1995.

GAO Report to Congressional Requesters, IT supply chain: National security-related agencies need to better address risks, United States Government Accountability Office, March 23, 2012.

Index

A

acceptance, acquisition plan, 55–6
accreditation, 150, 164
acquirers, 80–7, 90–2, 96–104, 115, 167, 200; maximizing visibility, 173–4; protecting contextual supply chain environment, 177
acquisition infrastructure: agreement processes, 46–7; concept of need, 51–2; ISO/IEC 12207 42–5; preparation, 50–1; process, 48–50
Acquisition Innovation Management process (SA-CMM), 263–4
acquisition plan, 50, 74–6; acceptance, 55–6; concept of need, 50–2; contracts, 45–51, 77; execution, 55; preparation, 54; system requirements, 53–4
Acquisition Planning process (SA-CMM), 246–7
acquisition process, 13, 35–6; customer agreement monitoring, 21–2; procurement program initiation and planning, 14–6; product acceptance, 22; product requirements communication and bidding, 16; project closure, 23; source selection and contracting, 16–20; supplier considerations, 20–1
acquisition requirements, 50, 76; preparation, 56–61
Acquisition Risk Management level (SA-CMM), 258–9
advertisement: acquisition, 57–8
agreements: acquisition infrastructure, 46–7; monitoring, 60–1

analysis, 169, 181, 187, 194, 200, 208–9, 232–4, 237; causal, 213, 227; impact, 210, 213–6; operational change, 210, 235; retrospective, 213; risk, 225, 229, 236; risk analysis process, 190; threat, 199; trade-off, 212, 236
architecture: controls, 192; processes, 272
assessment: SCRM (supply chain risk management) process, 26–7
assessment reports: updating, 160
assurance, 168–175, 178–188, 194–200, 237, 272; documentation, 207; establishing infrastructure, 225–8; goals, 205; ICT (Information and Communications Technology) products, 7–11; incorporating conditions in specifications of requirements, 175; operational, 208, 211–3, 225–6; proactive, 205–7; processes, 211; product assurance managers, 230–3; security control selection minimum requirements, 128–131, 134–5; SOW (statement of work) 17; suppliers, 46, 67, 72–4, 77; technical, 208
auditing operational supply chain system, 180
authentication, 28
authorization, 28; management, 212–6; response, 216–9; security controls, 149–155
authorization package, 142, 150–3, 164
availability, 39

B

baseline controls: NIST SP 800–53 catalog of, 190–1
baseline security, 272; control selection, 128–131

best practices, 82, 88–91, 99, 112–5, 170–2,
182–3, 194–200; controls, 96–7;
promoting trust, 92–3; SCRM
(supply chain risk management)
process, 172–181
BPR (Business Process Reengineering) 52
business process, 208
Business Process Reengineering (BPR) 52

C

capability maturity model (CMMs) *see* CMMs
(capability maturity models)
causal analysis, 213, 227
certification, 119, 150–2, 165; Federal
Information Security Management
Act (FISMA) certification, 119
change process, 219
changes: monitoring, 220–1
closure: acquisition requirements, 61
CMMs (capability maturity models) 240;
benefits, 241–2; *see also* SA-CMMs
commercial- off-the-shelf (COTS) system
security, 4, 15, 118
community of practice, 79, 115, 171, 200
compensation: security control selection, 131–2
compliance process, 211–2
compromise: supply chain, 9, 39
concept of need: acquisition, 50–2
confidentiality, 28, 39; supply chain
components, 174–5
configuration management, 44, 59, 77, 124,
139–142, 149, 158, 162, 165, 178–9,
184–5, 200, 237
consumers, 82–3, 88, 115, 200
continuity process, 212
continuous monitoring, 117, 125, 129,
133, 152, 165; accessing selected
security controls, 159; assessment
report updates, 160; determining
security impact of changes,
158–9; developing strategy,
136–8; implementing ICT system
decommissioning strategy, 162;
POA&M 160; remediation
actions, 159–160; reporting
security status, 160–1; reviewing
security status, 161–2; security
control selection, 136–7; security
plan update, 160; supply chain
risk, 155–7

Continuous Process Improvement process
(SA-CMM), 262
Contract Performance Management process
(SA-CMM), 257–8
Contract Tracking and Oversight process
(SA-CMM), 250
contracts, 18, 45–51, 65–7, 77; execution,
67–74; factors, 19–20; joint review
process, 59, 68–70, 74, 77
control assessment, 134, 137–9, 165; supply
chain security, 142–9
control framework, 96, 114–5, 200, 237, 272
control processes, 228–230
controls, 32, 35–9, 80–96, 99–100, 103–8,
115, 168–172, 179–186, 200, 237,
272; architecture, 192; best practices,
96–7; details, 187–9; enhancements,
193; implementing management,
191; limitations, 194–5; NIST SP
800–53 catalog of baseline, 190–1;
security, 119–137, 142–155, 186,
272; security control assessment
report, 165; statements, 192–3;
sustainment process, 203–234
COTS (commercial- off-the-shelf) system
security, 4, 15, 118
counterfeit ICT products, 5–6, 34, 39
countermeasures, 12, 39
customer agreement monitoring: ICT
acquisition, 21–2
customers *see* acquirers

D

day-to-day operational response process, 230–1
decision process, 213
decommissioning strategy: implementing ICT
162
design: defensive component development, 176–
7; process, 201, 213, 237; SCRM
(supply chain risk management)
process, 27–8, 37
design document, 27
development process, 205
discovery process, 216–217
disposal: reducing risk during, 182
diverse supply chain, 176
documentation: acceptance strategy, 55–6;
assurance, 207; security control
selection, 128; security controls,
141–2

E

environmental monitoring, 210
evaluation, 272
Evaluation process (SA-CMM), 251
exploration process, 217

F

factors: contracts, 19–20
Federal Information Security Management Act
(FISMA) certification, 119
FIPS 200 model, 182–6
First Principles, 108
FISMA 182
formal models: benefits, 170; building
processes, 167–170

G

generic security controls, 186
globalization, 110
governance: information, 39; organizational,
39; supply chain, 24–38

H

hardening supply chain delivery mechanisms,
179–180
HIPAA (Health Information Portability and
Accountability Act) 182
Humphrey, Watts, 8, 240

I

ICT (Information and Communications
Technology) products, 1; acquisition
process, 13–6, 21–3, 35–6;
breakdowns in supply chain, 6;
counterfeit, 5–6, 39; evolution, 2–3;
malicious logic, 5; procurement
program initiation and planning, 33;
product assurance, 7–8; supply chain
governance, 24–31; unintentional
vulnerabilities, 6; visibility, 9–12
impact analysis, 210, 213–6
incident response process, 181, 184–5, 200, 208
Information and Communication Technology
(ICT) *see* ICT (Information and
Communications Technology)
products

information governance, 39
infrastructure, 188, 200
Initial level (SA-CMM), 244
integration: SCRM (supply chain risk
management) process, 28–9
integrity, 29, 39
ISO/IEC 12207 acquisition infrastructure, 42–5

J

joint review process: contract agreements, 59,
68–70, 74, 77

K

key practices, 272
key processes (SA-CMMs) 242–4,
272; Acquisition Innovation
Management, 263–4; Acquisition
Planning, 246–7; Acquisition Risk
Management, 258–9; Continuous
Process Improvement, 262–3;
Contract Performance Management,
257–8; Contract Tracking and
Oversight, 250; Evaluation, 251;
Process Definition and Maintenance,
255–6; Project Management,
249–250; Project Performance
Management, 257; Quantitative
Acquisition Management,
261; Quantitative Process
Management, 260–1; Requirements
Development and Management,
248–9; Solicitation, 247–8; Training
Program Management, 259–260;
Transition to Support, 252–3; User
Requirements, 256–7

L

levels: SA-CMMs (capability maturity models)
242–244; Defined, 253–260;
Initial, 244; Optimizing, 262–4;
Quantitative, 260–1; Repeatable,
244–253
life cycles, 12

M

maintenance agreements: formalizing, 177–8
malicious code, 5, 34, 39

malicious logic: ICT (Information and
Communications Technology)
products, 5, 34
management authorization, 212–6
management controls: implementation, 191
management process, 210, 228
management roles, 230
maturity: processes, 264–6, 272; rating
schemes, 266–7
measure of confidence, 134
measurement: SCRM (supply chain risk
management) process, 31, 38
models: CMMs (capability maturity models)
240–2; FIPS 200 model, 182–6;
formal, 167–170; NIST SP 800–
53(Rev 4) model, 182–6; SA-CMMs
(capability maturity models)
242–264
monitoring, 180, 200, 208, 228–9, 237;
agreements, 60–1; changes, 220–1;
continuous, 117, 125, 129, 133,
152–165; environmental, 210; supply
chain vulnerabilities, 181

N

NIST (National Institute of Standards and
Technology) 9, 26, 39
NIST SP 800–53 baseline controls, 190–1;
feasibility, 188–9; limitations,
194–5
NIST SP 800–53 management controls,
191
NIST SP 800–53(Rev 4) model, 182–6
nonrepudiation of origin, 29

O

operational assurance, 208, 211–3;
methodologies, 225–6
operational change analysis, 210, 235
operational process, 209
operational response process, 230–1
opportunity identification: supply process, 63
Optimizing process (SA-CMM), 262–4
organizational governance, 39
organizational infrastructure, 188, 200, 225–7,
237
organizational parameters: security control
selection, 132
outsourcing, 201
oversight: suppliers, 72, 77

P

patches: software, 181
POA&M: updating, 160
proactive assurance, 205–7
Process Definition and Maintenance process
(SA-CMM), 254–6
processes, 172, 201, 227, 237; architecture,
272; assurance, 211; building with
formal models, 167–170; business,
208; change, 219; common features,
272; compliance, 211–2; continuity,
212; control, 228–230; decision, 213;
design, 201, 213, 237; development,
205; discovery, 216–7; exploration,
217; incident response, 208; key,
272; management, 210, 228;
maturity, 264–7, 272; operational,
209; operational response, 230–1;
protection, 176; reintegration, 223–5;
reporting, 211; response management
planning, 231–2; review, 208; SDLC
(system development life cycle) 272;
security information sharing, 228;
sourcing, 205–6
sustainment, 203–5, 212–234; testing,
220–2; violation, 208; vulnerability
response, 211
procurement program initiation and planning:
ICT acquisition, 14–6
producers, 229
product acceptance: ICT (Information and
Communications Technology)
acquisition, 22
product assurance: ICT (Information and
Communications Technology)
products, 7–11
product assurance managers, 230–3
product delivery and support, 74–5
product requirements communication and
bidding: ICT acquisition, 16
project closure: ICT acquisition, 23
Project Management (SA-CMM Repeatable
level) 249–250
Project Performance Management process
(SA-CMM), 257

Q

qualitative causal analysis, 213
Quantitative Acquisition Management process
(SA-CMM), 261

quantitative causal analysis, 213
Quantitative level (SA-CMM), 260–1
Quantitative Process Management level
    (SA-CMM), 260–1

R

rating schemes: process maturity, 266–7
reintegration process, 223–5
remediation actions, 216–9; continuous
    monitoring, 159–160
Repeatable level (SA-CMM), 244–6, 253–5;
    Acquisition Planning, 246–7;
    Acquisition Risk Management,
    258–9; Contract Performance
    Management, 257–8; Evaluation,
    251; Process Definition and
    Maintenance, 254–6; Project
    Management, 249–250; Project
    Performance Management, 257;
    Requirements Development and
    Management, 248–9; Solicitation,
    247–8; Transition to Support,
    252–3; User Requirements, 256–7
reporting process, 211
repositories, 213, 223, 230–2, 237
request for proposal (RFP) document, 16–8
Requirements Development and Management
    (SA-CMM Repeatable level) 248–9
requirements management, 180–1
requirements specification, 50, 77
research and development: costs, 15
response management process planning, 231–2
retrospective analysis, 213
review process, 208
RFP (request for proposal document)
    document, 16–8, 56–61
risk, 201, 237; disposal, 182, supply chain
    continuous monitoring, 155–7
risk analysis process, 190, 225, 229, 236
risk assessment, 123–5, 133, 143, 147, 153–4,
    157, 165, 272
risk evaluation: SCRM (supply chain risk
    management) process, 27
risk issues, ICT (Information and
    Communications Technology)
    products, 4–7
risk management, 117–9; procurement process,
    11; Risk Management Framework
    (RMF) 119; supply chain, 25–9,
    31–8, 170–2
risk mitigation, 152, 157, 165, 272

risk response, 237
RMF (Risk Management Framework) 119

S

SA-CMMs (capability maturity models)
    levels, 242–4; Defined, 253–260;
    Initial, 244; Optimizing, 262–4;
    Quantitative, 260–1; Repeatable,
    244–253
Saltzer, Jerome, 108
Sarbanes–Oxley Act (SOX) 182
SCDL testing, 178
Schroeder, Michael, 108
scoping: SCRM (supply chain risk
    management) process, 26, 36
SCRM (supply chain risk management) process,
    1, 24–31, 201; assessment, 26–7,
    36–7; best practices, 172–181;
    design, 27–8, 37; integration, 28–9;
    measurement, 31, 38; risk evaluation,
    27; risk management, 25; scoping, 26,
    36; standard model of best practice,
    32–3; threat identification, 26
SDLC (system development life cycle) 272
security assurance, 46, 67, 72, 74, 77
security control assessment plan, 165
security control assessment report, 165
security controls, 272; assessment,
    142–9; authorization, 149–155;
    categorization, 119–124;
    documentation, 141–2; generic, 186;
    implementation, 137–141; selection,
    124–135
security information sharing process, 228
security plans, 120–5, 128–133, 165; security
    control selection, 135–6; updating,
    160
service agreements: formalizing, 177–8
service delivery and support, 74–5
situational awareness: sustainment process,
    205–9
software updates, 181
Solicitation process (SA-CMM), 247–8
source selection and contracting: ICT
    acquisition, 16–20
sourcing, 201, 204–5, 234, 237; ICT
    acquisition, 13, 24–5, 31, 205–6
SOW (statement of work): assurance
    requirements, 17
specifications of requirements: supply chain
    assurance, 175

SRS (System Requirements Specification)
    document, 52–53
standard model of best practice: SCRM (supply
    chain risk management) process,
    32–-3
statement of work (SOW): assurance
    requirements, 17
statements: controls, 192–3
strategic planning, 28, 40
subcontractors, 229
supplementation: security control selection,
    132–3
supplier considerations: ICT acquisition, 20–1
supplier tendering, 63–5
suppliers: assurance, 46, 67, 72–74; oversight,
    72, 77; selection, 58–60; supply
    process, 61–74
supply chain, 159; auditing operational, 180;
    breakdowns, 6, 34; compromise,
    9, 39; confidentiality, 174–5;
    contextual environment protection,
    177; continuous monitoring risk,
    155–7; control assessment, 142–9;
    diversity, 176; establishing assurance
    infrastructure, 225–8; governance,
    24–37; hardening delivery
    mechanisms, 179–180; monitoring
    vulnerabilities, 181; reducing risks
    during disposal, 182; responding to
    incidents, 181; risk management,
    24–37, 170–2; security control
    assessment, 142–9; security control
    authorization, 149–155; security
    control categorization, 119–124;
    security control implementation,
    137–141; security control selection,
    124–136; weakest link, 9, 109–110
supply chain risk management (SCRM)
    process *see* SCRM (supply chain risk
    management) process
supply process, 61–2; contract agreements,
    65–7; contract execution, 67–74;
    opportunity identification, 63;
    supplier tendering, 63–5
sustainment process: controls, 203–5;
    authorized response, 216–9; building
    function, 228–9; document integrity,
    234; enforcing management control,
    233; generic management roles, 230;
    management authorization, 212–6;
    operational impacts, 209–212;
    operational response process, 231–2;

reported vulnerability analysis,
    209–212; response evaluation,
    220–3; response integration, 223–5;
    situational awareness, 205–9; status
    assessment, 233–4
system development life cycle (SDLC) *see* SDLC
    (system development life cycle)
system requirements: acquisition, 53–4
System Requirements Specification (SRS)
    document, 52–3

T

technical assurance, 208
testing, 201, 238, 272; processes, 220–2; SCDL
    178
threat analysis, 199
trade-off analysis, 212, 236
Training Program Management level
    (SA-CMM), 259–260
Transition to Support process (SA-CMM),
    252–3
transparency, 201, 238
trust: best practices, 92–3

U

unintentional vulnerabilities: ICT (Information
    and Communications Technology)
    products, 6
updating software, 181
User Requirements process (SA-CMM), 256–7

V

validation, 44, 60–1, 66–9, 72–3, 76–7, 130,
    139–143, 165, 201, 222, 238
verification, 44, 60–1, 66–9, 72–3, 76–7, 139,
    143, 149, 163–5, 201, 213, 222, 229,
    238
violation processes, 208
visibility: acquirers, 173–4; ICT (Information
    and Communications Technology)
    products, 9–12
vulnerabilities, 272; discovery process, 216–7;
    supply chain monitoring, 181
vulnerability response management, 209–212
vulnerability response process: fixes, 217–9

W

weakest link: supply chain, 9, 109–110

Printed and bound by CPI Group (UK) Ltd, Croydon, CR0 4YY

21/10/2024

01777106-0001